£1-70
(12)

MEMORIES

OF THE

L.N.E.R.

in

Rural Northumberland

2nd (Enlarged) Edition

by

Allan W. Stobbs, M.S., B.Sc.

Published by the author at
Caroline House, Lowther Street,
Penrith, Cumbria, CA11 7UW.

Published 1986
Reprinted 1988
This Edition 1992

ISBN 0 9515330 1 0

Printed in Great Britain by
Reed's Limited, Penrith, Cumbria.

Contents

Map of lines covered in this book

Maps and Diagrams

Illustrations

Foreword to the 1992 edition

Since *Memories of the LNER - Rural Northumberland* was first published in 1986, I have received a considerable amount of information by letter and telephone from people who had direct experiences with these railways.

In the first edition I emphasised that what I had written was about my own acquaintance with the railways mentioned and the memories that they evoked. I did not copy or repeat what had been said or written by anyone else other than historical facts.

In this edition I have included information and experiences passed on to me by a number of well-wishers who are willing to add to the fund of historical knowledge relating to the rural railways of Northumberland. Therefore I can do no better than repeat the words of Tom Lehrer the Harvard mathematics teacher turned satirical song writer and singer. He made the point that if you appropriate the thoughts and ideas of one person and then present them as your own, you are guilty of the degrading sin of plagiarism. If, however, you appropriate the thoughts and ideas of a number of people then you are engaged in the highly commendable activity of research!

It is with the full knowledge and approval of the persons concerned that I include my "research" findings in this edition.

It is always invidious to mention names but of the many people who have sent me information, and to all of whom I am extremely grateful, I wish to record my special thanks to Mr. Ken Appleby for correcting my errors and providing much additional information on signals and signalling. Dr. J. L. Middlemas, a former G.P. at Bellingham, has presented the human side of the Border Counties line and the part that it played in the welfare of families living in remote situations to the North Tyne valley. Mr. J. B. Dawson and Mr. Alex Appleton have both contributed most interesting details of operations and activities of which I was hitherto unaware. Finally, Mr. Bill Lynn has provided a wealth of information on locomotives and operating practices on the Border Counties, Rothbury and Wanney lines. He has given me much more material than I have been able to use. To all these people my thanks.

After the publication of the first edition I received some criticism because no mention had been made of the Alston and Allendale lines. The reason for their exclusion was that since I had not travelled on either of them in L.N. E. R. days nor had I been associated with them in any way, then they did not qualify under the criteria that I had set out to honour the title.

A new chapter has therefore been added in this edition to make good the omission. It is titled "The Lead Lines" and it deals in some detail with both the Alston and Allendale lines, so completing the record of all the principal branch lines operated by the L.N.E.R. in rural Northumberland.

Introduction

When we were young, the world around us seemed to be a well-ordered and stable place. Things happened with measured regularity at the same time, at the same place and in the same way. Like clockwork. The same events took place daily or weekly, as the case may be, everyday and every week. They always did, they always had done and we assumed that they always would.

As children, we were awakened at half past seven every morning by the shipyard buzzahs. They told us when it was dinner-time at twelve noon and when the hour was up at one o'clock. They announced the end of the working day at five o'clock when thousands of weary men would stream from Smiths Dock, Swan Hunters, North Eastern Marine, Walker Naval Yard and a host of other industrious workplaces.

Every Saturday between eleven and twelve o'clock, a procession of tram-cars trundelled along Market Street, Newcastle on their way to Delaval in time for workmen pouring out of Vicker-Armstrong's Scotswood and Elswick works.

Every "Blue" bus of Newcastle Corporation Transport carried a blue identification light alongside the destination blind at the front of the bus. This was to distinguish it from vehicles of other operators.

Every week the tall-funnelled S.S. "Jupiter" and the M.V. "Venus" sailed from the Tyne Commission Quay at Percy Main to Bergen and Stavanger.

These, and lots of other things that happened in our world, were regular and reliable and would go on forever – or so we thought.

It was only as we grew older that we began to realise that the environment in which we lived was not stable or static, that some things did change and that the older we became the faster the changes occurred. In no sphere of activity has this been more pronounced than on the railway.

What a traumatic experience it must have been for railwaymen of the old constituent companies when the great amalgamation took place in 1923. Old loyalties, built up over a lifetime of service, were shattered overnight. Rival companies that had fought bitterly over the years for routes and traffic suddenly became colleagues in the new organisation. Long established working practices were abandoned and changed and familiar signs and symbols were replaced by new alien ones.

In Northumberland the rival North Eastern and North British Railways became part of the London and North Eastern Railway. Yet it was not until 1945 that the ex-North British lines in the county were integrated into the North Eastern Area of the LNER. Before that year they had been omitted from North Eastern area time-tables and public services were advertised on the same wall time-table as the Edinburgh to Carlisle "Waverley" route and branch lines to Eyemouth and Silloth.

The Wooler line, on the other hand, being ex-North Eastern, was fully integrated into the North Eastern area time-table even though there were no passenger services.

The anomaly was further compounded by the fact that the locomotive department of the LNER regarded the small engine shed at Rothbury as a sub-shed of Heaton. It was supplied with two ex-North Eastern locomotives, an 0-4-4 tank engine (LNER class G5) and an 0-6-0 goods engine (LNER class J21) and these were serviced and repaired at Heaton. Reedsmouth, to be different, had a stock of ex-North British locomotives, mainly 0-6-0 tender engines (LNER class J36) and they were sent to Cowlairs for major servicing and repairs. This state of affairs lasted throughout the life of the LNER.

Tremendous changes have taken place everywhere during the last fifty years. The North-east has had its fair share, some say much more than its fair share, of change. Coal mines, steelworks and shipyards have either disappeared or are struggling to survive. The railways have been mutilated.

In the 1930s, the LNER operated 316 miles of railway in Northumberland and there were 119 stations open for passengers. In 1992, British Rail operate over only 130 route miles with 20 stations open for passengers, most of them unstaffed.

A large proportion of today's Tyneside population has no idea as to how busy Newcastle Central station was in the 1930s. Platforms 11-15 at the west end were in constant use by trains either departing for or arriving from Carlisle and Hexham with through carriages for Stranraer; Blackhill via Birtley; Blackhill via Scotswood; Hawick and Border Counties line stations; and North Wylam. At the other end of the station, platforms 1-7 were used by trains on the North and South Tyneside electric lines and by stopping trains to Alnwick; Sunderland and Middlesbrough; and Ferryhill and Leamside. Platform 4 was a long one, running right back into the main square, and it was used for Manchester and Liverpool trains routed by the old Leeds Northern line via Sunderland, Stockton, Northallerton, and Ripon to Leeds. Half a mile east, at Manors North, passenger trains departed for stations on the former Blyth and Tyne Railway to Blyth, Ashington, Newbiggin, and Morpeth via Bedlington.

Now, apart from Inter-city services, Newcastle Central provides trains to only Sunderland and Middlesbrough, Hexham and Carlisle, and a local service to Morpeth, Alnmouth and Berwick. Manors North has ceased to exist.

All that now remains of these abandoned railways are some remnants of the infrastructure and the memories that they created. But memory is a fickle faculty and I would not guarantee that everything in the pages that follow is wholly accurate. They have been recorded as I remember them in the hope that they may recall similar memories to readers of the older generation. Perhaps they may also acquaint younger readers with the fact that it was possible, not only to survive, but to live full and active lives before universal dependence on the little tin boxes on four rubber wheels!

It is still possible to trace the routes of these now forgotten railways by structures that have been left behind, structures for which there is no second-hand market and no realisation of invested capital. I refer, of course, to cuttings, embankments, tunnels and bridges which involved a lot of civil engineering work and many, many man-days of hard physical labour. These can be found relatively easily in open country but in urban areas they have mostly disappeared.

Some stations, as at Humshaugh, have been converted into delightful dwellings; some, as at Wooler and Scot's Gap have found alternative industrial uses; while others, like Whittingham, have just been allowed to deteriorate and become something of an eyesore. Rothbury station has disappeared without trace and replaced by a so-called industrial estate, although it is doubtful if it employs as many people as the railway once did.

Northumberland is essentially a livestock county. It is highly regarded both for the number and quality of cattle and sheep that are bred and reared within its borders. Before the Railway Age, livestock had to walk, often great distances, from their place of birth to the areas where they were fattened and slaughtered. This was the period of the drove roads when livestock were bought and sold at fairs held periodically in market towns and often associated with Saints Days. This was the normal method of livestock trading in country towns such as Berwick, Alnwick, Morpeth and Hexham. The decline of livestock fairs can be attributed to railways for they were responsible for the establishment of livestock auction markets, or marts. It is no coincidence that marts were sited adjacent to stations because the railway was the raison d'etre of the mart and the prime mover of cattle and sheep after they had been sold.

A study of station diagrams shows that the mart and the railway were interdependent at Wooler, Alnwick, Rothbury and Scot's Gap. The same was also true at Bellingham, Acklington, Belford and Haltwhistle. Only at Hexham were the two marts some distance from the station and, when evacuated to Hexham, I spent many days herding cattle from Tynedale mart down to the cattle dock at the goods yard.

The care and welfare of livestock carried by train was of great concern to the railway companies and their staffs. It was far superior to that provided by road hauliers today. The railways were obliged, by law, to water all livestock every four hours and cattle sidings were scattered throughout the country for this purpose. Livestock trains would be drawn into these sidings and, wagon by wagon, cattle and sheep would be let out on to the penned platform side to drink from long water troughs. Many animals today suffer long journeys in road transporters in hot weather without any provision being made for them to get water. Driver's hours and the mileage covered in a given time are much more important than animal welfare. These facts are apparently not appreciated or recognised by the animal welfare lobby for it has made no attempt to have the high standards of livestock movement by train made mandatory for livestock carried by road vehicles. So much for "fair" competition.

It will be seen, then, that railways were largely responsible for the growth and development of Northumberland's small towns as livestock centres plus all the ancillary activities associated with a thriving agriculture. Having accepted all the benefits that the railway brought, these same towns have now cast the railway aside with negligible gratitude or protest.

No longer does the 10-13 a.m. from Riccarton pick up two wagon loads of cattle at Kielder, as I have seen it do, and take them to Bellingham for the weekly cattle sale – a distance of 16 miles! We have progressed since then. Times have changed. The safety of the railways has been exchanged for the slaughter on the roads. It will be left to historians in the years ahead to decide whether or not these changes have been for the betterment of mankind.

Meanwhile, the memories linger on.

Photo: John Daysh

One of the un-named Class D32 4-4-0's No. 2449 at Corbridge with a train for the Border Counties Line in the 1940s.

Photo: R. W. Lynn Collection

Border Counties Junction, one mile west of Hexham. The lines to the right lead on to the single track bridge over the Tyne. This bridge was badly damaged in 1948 and ultimately became the reason for the line's closure. The Allendale line came in from the left immediately beyond the signal post.

Photo: Ian S. Carr

Class D49/1 4-4-0 No. 62733 "Northumberland" with Thompson corridor stock joins the Carlisle-Newcastle line at Border Counties Junction on 29th August 1956. Note the 1950s repair to the bridge which so reduced clearances that locos such as the D30's could no longer be used.

The Borders Counties Line

L.N.E.R. branch lines in Northumberland included 75 miles of the former North British Railway. The incursion south of the border by this Scottish company had been rather like one of the old Border Raids. It advanced discreetly through a little known pass in the hills from Riccarton (on the Waverley route) to Deadwater at the head of the North Tyne valley and then followed the river down to Hexham. Chronologically it wasn't quite like that but the intent was the same. The aim of the North British was to break the monopoly enjoyed by the North Eastern Railway on all traffic between Edinburgh and Newcastle. By 1865 the North British had succeeded in getting as far as Morpeth in the east and Hexham in the south. Their ambition of getting their own route into Newcastle from Morpeth was dashed when the Blyth and Tyne Railway was absorbed by the North Eastern in 1874. The best that they could do at Hexham was to get running powers over the North Eastern from Border Counties Junction to Newcastle and the right to build an engine shed and goods station at Newcastle. These facilities, however, were only conceded if they (the North British) gave running powers to the North Eastern from Tweedmouth to Edinburgh.

Even though the former North British lines in Northumberland became part of the LNER in 1923, the passenger timetables continued to be published as part of the North British section (on the Edinburgh to Carlisle sheet) and ex-North British locomotives used on the lines were sent to Cowlairs for repairs and major overhauls. While the North Eastern section was responsible for providing two 3-carriage sets for the Newcastle-Hawick service, the North British section (Southern Scottish area) was responsible for carriages on the Rothbury and Wanney lines.

Strangely enough, my first "memory" of the Border Counties railway is one that I do not remember! At the age of two I was taken on holiday to Bellingham where we stayed in a house near to the station. Apparently the entire household was awakened at 7-28 every morning by my ecstatic shrieks as the first train of the day from Riccarton arrived at Bellingham. The sound of the safety valves blowing off and the starting whistle were obviously music to my ears! This was followed by an "action replay" twenty minutes later when the first north-bound train arrived from Hexham. This is a memory which, although I cannot recall it, I have nevertheless never been allowed fo forget.

Passenger traffic on the 42-mile line was always very light. There were three trains a day in each direction on Mondays to Saturdays with an additional "shopping" train on Saturdays. This varied. In 1910, for example, the North British ran a train from Bellingham to Hexham at 2-15 p.m. and it returned from Hexham at 7-00 p.m. In 1938 the LNER ran a train from Kielder to Hexham at 1-29 p.m. which didn't return from Hexham until 10-25 p.m., arriving at Kielder at 11-53 p.m. when the empty stock was returned to Reedsmouth. From 1944 the service ran to Hexham only on alternate Saturdays and involved empty stock being taken from Reedsmouth to Kielder in the morning and then being returned at night. The other Saturdays had a late train from Hawick at 10-45 p.m. which ventured as far as Falstone in order to cater for passengers at Lewiefield Halt. It arrived at Falstone at 11-55 p.m. Empty stock was then returned to Riccarton and stabled in sidings at Stobs Camp where it arrived at two o'clock on Sunday morning. The engine, usually a D30 4-4-0 running tender first, then returned to Hawick. Apparently this train was known locally as the "Kilty" because most of the passengers were from Glasgow. By 1956, when the line was closed, the extra Saturday train ran from Kielder to Hexham and back again with empty carriages taken out from Hexham in the morning and brought back again at night by a Hexham based V1/V3 2-6-2 tank engine.

The everyday service in 1944 consisted of an early morning train which left Hawick at 6-05 a.m., called at Riccarton at 6-37 a.m., arrived at Hexham at 8-24 a.m. and Newcastle at 9-05 a.m. The second south-bound train commenced at Riccarton at 10-20 a.m., arrived at Hexham at 11-55 a.m. and then ran express to Newcastle, calling only at Stocksfield, to arrive at Newcastle at 12-36 p.m. The third train began at Hawick at 4-25 p.m., called at Riccarton at 4-58 p.m. and terminated at Hexham at 6-28 p.m.

In the opposite direction the first train of the day left Newcastle at 5-55 a.m. and called at all stations to Hexham, arriving there at 6-51 a.m. It terminated at Riccarton at 8-51 a.m. The other two north-bound trains ran through to Hawick. One left Newcastle at 11-10 a.m. and Hexham at 12-10 p.m. arriving at Riccarton at 1-50 p.m. and Hawick at 2-18 p.m. The other one began at Hexham at 5-10 p.m., arrived at Riccarton at 6-45 p.m. and Hawick at 7-19 p.m. This was the general pattern of services over the years, although times varied by a few minutes and so did the station from which they began and ended their journeys. In 1922, for instance, the last train north began at Newcastle at 4-27 p.m., called at Hexham at 5-08 p.m., arrived at Riccarton at 6-49 p.m., Hawick at 7-12 p.m. and finally terminated at Edinburgh at 9-01 p.m. I imagine that none but North British pass holders would choose to spend 4 hours and 26 minutes in travelling from Newcastle to Edinburgh by that train.

When the line was built there were great hopes that outcrop coal at Plashetts and Bellingham would provide considerable mineral traffic. The colliery at Plashetts operated successfully until 1926 and was relatively busy but coal mining at Bellingham failed to develop on any scale. Bellingham also had an

RICCARTON to HEXHAM and NEWCASTLE.

UP TRAINS.		Distance from Riccarton.		WEEK-DAYS. 3	4	5	6	7	8	10	11	12	13	14	15	16	17	18	19	SUNDAYS. 1	2
				O.P.	Goods		O.P.	O.P.	Goods	O.P		Goods	O.P.	O.P.		O.P.					
CLASS				...	D	...	...	...	D		...	B	...	...	...	...	...	...	...	...	...
Departs from				Hawick a.m. 6 5	...	...	...		Hexham a.m. 8 10	...	...	Hawick p.m. 12 45	...	Hawick p.m. 4 25	...	Hawick p.m. 10 45	...	...	...	...	...
		M.	C.	Z a.m.	a.m.		a.m.	p.m.	a.m.	‡ p.m.		p.m.	p.m.	V p.m.		SO p.m.					
—Riccarton Junction	...dep.	...	...	6 37	9 0	...	10 20	...	...	...	...	2 ‡ 8	...	4‡58	...	11 18	...	...	...	...	...
Saughtree	... "	2	46	6 41	*	...	10 24	...	...	...	...	...	...	5‡ 2	...	11 23	...	...	...	...	...
Deadwater	... "	5	52	6‖47	*	...	10 30	...	...	...	...	...	...	5‡ 8	...	11 30	...	...	...	...	...
—Kielder	arr.	8	18	6 52	...	...	...	...	...	...	...	...	...	...	...	...	...	...	...	...	...
	dep.	...	...	6 53	*	...	10 35	...	...	1 33	...	2 27	...	5‡14	...	11 36	...	...	...	...	...
Lewiefield (Halt) ...	... "	10	10	6 58	...	...	10 40	...	...	1 40	...	...	...	5‡19	...	11 43	...	...	...	...	...
Plashetts	arr.	11	50	‡	...	...	10 43	...	...	...	...	...	...	5‡22	...	...	...	...	...	...	...
	dep.	...	...	‡	*	...	10 44	...	...	1 46	...	...	...	5‡23	...	11 48	...	...	...	...	...
—Falstone	arr.	16	70	7 9	...	...	10 51	...	...	...	...	...	...	5‡30	...	11 55	...	...	...	...	...
	dep.	...	...	7 10	*	...	10 52	...	...	1 55	...	2 48	...	5‡31	...	...	...	...	...	...	...
Thorneyburn ...	... "	20	15	7 17	...	...	...	...	...	2 2	...	...	...	...	...	Runs on alternate Saturdays, commencing 3rd June, 1944.	...	...	...	...	...
Tarset	... "	21	40	7 21	*	...	11 1	...	...	2 6	...	...	...	5‡40	...		...	...	...	...	...
—Bellingham	arr.	25	6	7 27	...	...	11 7	...	...	...	...	...	...	5‡46	...		...	...	...	...	...
(North Tyne)	dep.	...	...	7 28	*	...	11 8	...	...	2 15	...	...	4 5	5‡47	...		...	...	...	...	...
—Reedsmouth Junction	arr.	26	65	7 32	*	...	11 12	...	...	...	...	3 10	4 9	5‡51	...		...	...	...	...	...
	dep.	...	...	7 48	...	...	11 23	...	*	2 20	...	3 20	4 15	5 56	...		...	...	...	...	...
—Wark	arr.	30	56	7 55	...	...	11 30	...	*	...	...	...	...	6 3	...		...	...	...	...	...
	dep.	...	...	7 57	...	...	11 31	...	*	2 28	...	...	...	6 4	...		...	...	...	...	...
Gunnerton Siding ...	... "	33	0	...	...	...	...	...	...	...	...	...	...	...	...		...	...	...	...	...
Barkus' Siding ...	... "	33	55	...	Min.	...	...	...	*	...	...	...	...	...	...		...	...	...	...	...
Barrasford	... "	34	38	8 3	CL D	...	11 37	...	*	2 35	...	...	...	6 10	...		...	...	...	...	...
Chollerton	... "	35	54	8 8	Blaydon 5a20	...	11 40	...	*	2 38	...	...	...	6 13	...		...	...	...	...	...
Cocklaw Siding ...	... "	36	30	...		...	...	...	*	...	...	...	...	...	...		...	...	...	...	...
Humshaugh ...	... "	37	2	8 12		...	11 44	...	*	2 41	...	...	...	6 17	...		...	...	...	...	...
—Wall	arr.	38	67	8 15	a.m.	...	11 47	...	*	...	...	...	...	6 20	...		...	...	...	...	...
	dep.	...	...	8 17	*	...	11 48	...	*	2 46	...	3 50	...	6 21	...	...	...	...	...	...	...
Acomb Sidings ...	... "	40	19	...	...	...	...	...	...	...	...	...	...	...	...	...	...	...	...	...	...
—Border Counties Junc.	... "	40	71	8 22	...	...	11 53	...	...	2 51	...	4 0	...	6 26	...	...	...	...	...	...	...
N.E. Area. Hexham ...	arr.	41	78	8 24	DD	...	11 55		*	2 53	...	4 5	...	6 28	...	...	...	...	...	...	...
	dep.	41	78	8 28	...	...	12 8	12 3	Weigh and marshal Stone traffic at Wall.	‡ Runs on alternate Saturdays, commencing 27th May, 1944.	...	4 20	...	...	...	...	...	...	...	...	...
Corbridge ...	... "	45	7	...	...	...	12 14	...			...	4 27	...	...	...	...	...	...	...	...	...
Riding Mill ...	... "	47	31	—	...	...	12 19	...			...	...	...	...	...	...	...	...	...	...	...
Stocksfield ...	... "	49	55	...	...	...	12 27	12 15			...	...	...	...	...	...	...	...	...	...	...
Mickley ...	... "	51	26	...	...	...	...	...			...	...	...	...	...	...	...	...	...	...	...
Prudhoe ...	... "	52	21	...	...	...	12 33	...			...	...	...	...	...	...	...	...	...	...	...
Wylam ...	... "	54	32	...	...	...	12 38	...			...	...	...	...	...	...	...	...	...	...	...
Ryton	... "	56	50	...	...	...	12 43	...			...	...	...	...	...	...	...	...	...	...	...
Blaydon ...	... "	58	63	8 51	...	...	12 48	...			...	...	...	...	...	...	...	...	...	...	...
Scotswood ...	... "	59	60	8 55	...	...	12 53	...			...	...	...	...	...	...	...	...	...	...	...
Elswick ...	... "	61	0	9 0	...	...	12 57	...			...	...	...	...	...	...	...	...	...	...	...
Newcastle ...	... arr.	62	66	9 5	...	...	1 2	12 36			...	...	...	...	...	...	...	...	...	...	...
Arrives at		...	...	...	...	...	...	...	...	...	...	Hawick 8 42	Scots gap 4 48		...	Stobs Camp 1†15	...	...	...	...	...

No. 3.—‖ Calls at Deadwater, when required to take up, on application being made to the Station Master at Riccarton Jct. before 5.0 p.m. the day previous to travel. ‡ Calls at Plashetts at 7.1 a.m. on Tuesdays and Saturdays, when required, to pick up Passengers.

No. 10.—Reedsmouth to provide Engine, Guard and Vehicles.

No. 12.—‡ Leaves Riccarton 1.55 p.m. and runs correspondingly earlier when ready to do so.

No. 14.—‡ On Saturdays departs Riccarton Junction 5.0, Saughtree 5.4, Deadwater 5.10, Kielder 5.15, and runs one min. later to Reedsmouth arrive, thence as booked. Enginemen and Guard change at Reedsmouth with Men of No. 12 Down. N.E. Area Guard works Empty Carriages, leaving Newcastle at 7.35 p.m. and arriving Delavel Sidings 7.42 p.m.

B 25 m.p.h. D 20 m.p.h.

E—4c

From the L.N.E.R. (Southern Scottish Area) "Working Timetable", May-September 1944

ironworks but this was closed soon after the railway opened. However, considerable quantities of sintered iron ore came on to the line at Reedsmouth from Broomhope and there were large quantities of stone from quarries at Barrasford and Cocklaw. Travellers today on the military road from Heddon-on-the-wall to Greenhead will have great difficulty in believing as they look down on what was Humshaugh station, that at one time there was a colliery nearby and batteries of coke-ovens – but such was the case. Revenue-earning traffic during LNER days consisted of small quantities of coal from Plashetts, livestock to and from Bellingham Auction Mart (mainly during the period August to October) stone from Barrasford and Cocklaw and coal from Acomb colliery – 3 miles north of Hexham. This coal traffic continued up to the closure of the line and involved a daily train to Blaydon, invariably hauled by a J27 0-6-0.

Prior to 1923 the line was used intensively by the North British in order to get their goods traffic to Newcastle with the minimum use of North Eastern track. Use of the line was further intensified during the Great War (1914-1918) when it was open 24 hours a day. I have received a working time-table for July 1918 which shows that six goods trains a day ran between Scottish and Tyneside depots. The most famous of these was the "beer" train which ran from Duddingston, Edinburgh to the North British goods station at Forth Banks, Newcastle. Duddingston was right in the heart of old Edinburgh, to the east of St. Leonard's Street, and surrounded by the breweries of Deuchars, McEwan's and Younger's. Access was by a single track branch line which left the Morningside circle line at Craigmillar station and approached the goods station through a tunnel to avoid spoiling the view south-west from King's Park. The "beer" train left Duddingston at 9-10 p.m., passed Riccarton at 1-48 a.m., stopped at Reedsmouth for 20 minutes at 2-50 a.m., to take water, passed a north-bound goods to Meadows at Wark at 3-26 a.m. and arrived at Forth at 5-18 a.m. – having taken 8 hours and 8 minutes for the journey. It ran seven days a week but arrived at Forth 15 minutes later on Monday mornings. Although the advantage of this route to the North British disappeared with the formation of the LNER. I am told that the "beer" train continued to run well into the 1930s.

In 1918 there were three trains a day from South Leith and one from Grangemouth to Blaydon. These were class B goods and conveyed Admiralty coal empties for Wales and Durham collieries with the load restricted to 48 wagons. What a sight one of these trains must have been, snaking down the North Tyne valley. They left South Leith at 10-00 p.m., 2-35 a.m. and 5-30 a.m. and Grangemouth at 4-35 p.m. with arrivals at Blaydon at 5-28 a.m., 11-18 a.m., 12-43 p.m. and 2-45 p.m. The coal-burning warships of the Grand Fleet must have used prodigious quantities of coal, all of which had to be taken by train to Wick or Thurso for Scapa, Inverness for ships in the Moray Firth; Rosyth and Grangemouth. And, of course, the empties had to be returned to the collieries.

There was also a through goods train from Galashiels to Newcastle Forth. It left at 1-00 p.m. and called at all stations between Riccarton and Wall to pick-up and leave livestock and other urgent traffic. It arrived at Forth at 9-57 p.m. having taken 8 hours 57 minutes for 94 miles.

On Wednesdays, at this period, a stores train ran from Carlisle, at 6-45 a.m., reversed at Riccarton and arrived at Reedmouth at 10-20 a.m. It stayed there for an hour, left at 11-22 a.m., got to Riccarton for 12-10 p.m. and then proceeded north to St. Boswells.

In the reverse direction there were four goods trains a day (five on Saturdays) from West Dunston to Meadows (Edinburgh). The first one left at 1-45 a.m. and arrived at Meadows at 9-48 a.m. and the last one left West Dunston at 9-30 p.m. during the week but 11-05 p.m. on Saturdays and arrived at 6-20 a.m., and only 10 minutes later on Sunday morning.

The balancing Galashiels goods left Newcastle Forth at noon, called at Elswick, when required, to pick-up gun traffic for Broomhope, and at Blaydon and Mickley for coal before arriving at Hexham at 2-20 p.m. It then called at Wall for 4 minutes, Reedsmouth for 31 minutes where any gun traffic would be detached, and Kielder for 10 minutes before arriving at Riccarton at 5-35 p.m. and Galashiels at 8-28 p.m.

The remaining goods of the day went to Portobello and travelled through the night. It left Forth at 7-55 p.m., called at Elswick if there were any guns for Broomhope and then only at Blaydon to arrive at Hexham at 9-31 p.m. It stopped at only Reedsmouth (22 minutes) and Kielder (4 minutes) on the Border Counties line and arrived at Riccarton at 12-20 a.m. Arrival time at Portobello was 4-45 a.m.

In addition to through goods trains there were three local pick-up trains. One left Riccarton at 6-50 a.m. and called at all stations to Reedsmouth where it arrived at 10-12 p.m. It then made the trip to Broomhope siding or Woodburn when required. The return from Reedsmouth was at 1-30 p.m. It called at all stations except Deadwater and Saughtree and arrived back at Riccarton at 4-34 p.m.

At the south end of the line there was a train from Hexham at 7-35 a.m. which called at Wall, Chollerford (later renamed Humshaugh), Barrasford and Wark and arrived at Reedsmouth at 10-30 a.m. It also made a trip on the Wanney line, close on the heels of the train from Riccarton, and went through to Scots' Gap. It returned from there at 1-45 p.m. and left Reedsmouth at 3-30 p.m. calling at Wark, Barkus siding, Barrasford, Chollerton, Chollerford, Wall, North Tyne colliery and Howford siding to arrive at Hexham at 6-05 p.m. It was a real pick-up goods! A second goods left Hexham at 9-30 a.m., called at North Tyne colliery, Wall, Cocklaw quarry, Chollerton, Gunnerton siding and Wark to arrive at Reedsmouth at 11-28 a.m. This train is of interest because it did not return to Hexham. It was worked by the engine and men of the 8-00 a.m. passenger train from Bellingham to Hexham. Question? "How was the passenger train returned from Hexham to Bellingham"!

I have dealt in some detail with these goods train operations in order to illustrate the importance of the Border Counties line to the North British Railway as their main artery south from Edinburgh to Newcastle. In all, there were 11 trains a day each way which is not bad for a single track branch line where there were three sections eight miles long and took 20 minutes each to travel. Indeed, the Border Counties line probably carried more goods traffic in 1918 than the East Coast main line does today! It also leads to the question that since the Riccarton engine dealt with traffic from Reedsmouth to Broomhope and Woodburn, and a Hexham engine handled traffic to Scot's Gap, what did the four engines based at Reedsmouth do?

The creation of the LNER eliminated the need to run Edinburgh to Newcastle goods traffic over the Border Counties line but the line did serve a similar function in connection with meat traffic from Hawick to London. The obvious route was by Carlisle but, since this involved a relatively short mileage on LNER tracks and a large one on LMS lines, the Border Counties line was used. Therefore, on Tuesdays, a special meat train left Hawick at 1-00 p.m. and travelled over the Border Counties to arrive at Hexham at 4-50 p.m. Special arrangements were made for it to be attached to the 5-05 p.m. passenger from Hexham and this arrived at Newcastle at 5-54 p.m. The meat vans then went forward to King's Cross East Yard as part of an express meat train which had left Edinburgh at 2-53 p.m. Departure from Newcastle was at 6-17 p.m., thanks to some smart work by shunters and the Central station pilot engine. This train was due at King's Cross at 1-02 a.m. where the meat was trans-shipped to Smithfield market.

I have also been reminded of another interesting working which began, of all places, at Corbridge. This train ran daily in the 1940s to convey coal from Durham collieries to Hawick gasworks and it originated at Corbridge because there wasn't any available siding accommodation at Hexham. Empty wagons left Hawick begind a J36 0-6-0 at 12-45 p.m. and got to Riccarton about an hour later. The train left Riccarton any time between 1-55 p.m., and 2-08 p.m., stopped at Reedsmouth for 10 minutes, Hexham for 15 minutes and arrived at Corbridge at 4-27 p.m. Thirty-three minutes were allowed to exchange the 25 or so empty wagons for a similar number of full ones and these got back to Hexham for 5-08 p.m. The engine was then turned and watered and left at 5-20 p.m. It stood for 13 minutes at Wark to allow the 4-25 p.m. ex-Hawick to Newcastle passenger to pass, and 15 minutes at Reedsmouth – for no apparent reason – and arrived at Riccarton without further hindrance at 7-48 p.m. Hawick arrival time was 8-42 p.m.

Kielder station could have been one of the busiest on the line if Dr. Beeching and British Railways Board had shown any foresight as to what the development of Kielder Forest meant in terms of timber production. Established in 1926, it has grown to a size of more than 80,000 acres and has become the largest man-made forest in Europe. Allowing for an average timber harvest age of 50 years, this area will be producing more than 5,000 tons of timber a week by the year 2,000. This should all have been taken out of the North Tyne valley by train. Even by 1956, when the railway was closed, 42,000 acres had been planted in Kielder Forest alone with another 18,900 and 11,300 acres in the adjoining Wark and Redesdale Forests. Sadly and very misguidedly, the railway was closed at a time when the extraction of timber from trees planted in 1926 was just beginning to gain momentum. And all for the want of a paltry £28,000 to repair the storm damaged bridge over the Tyne at Border Counties Junction.

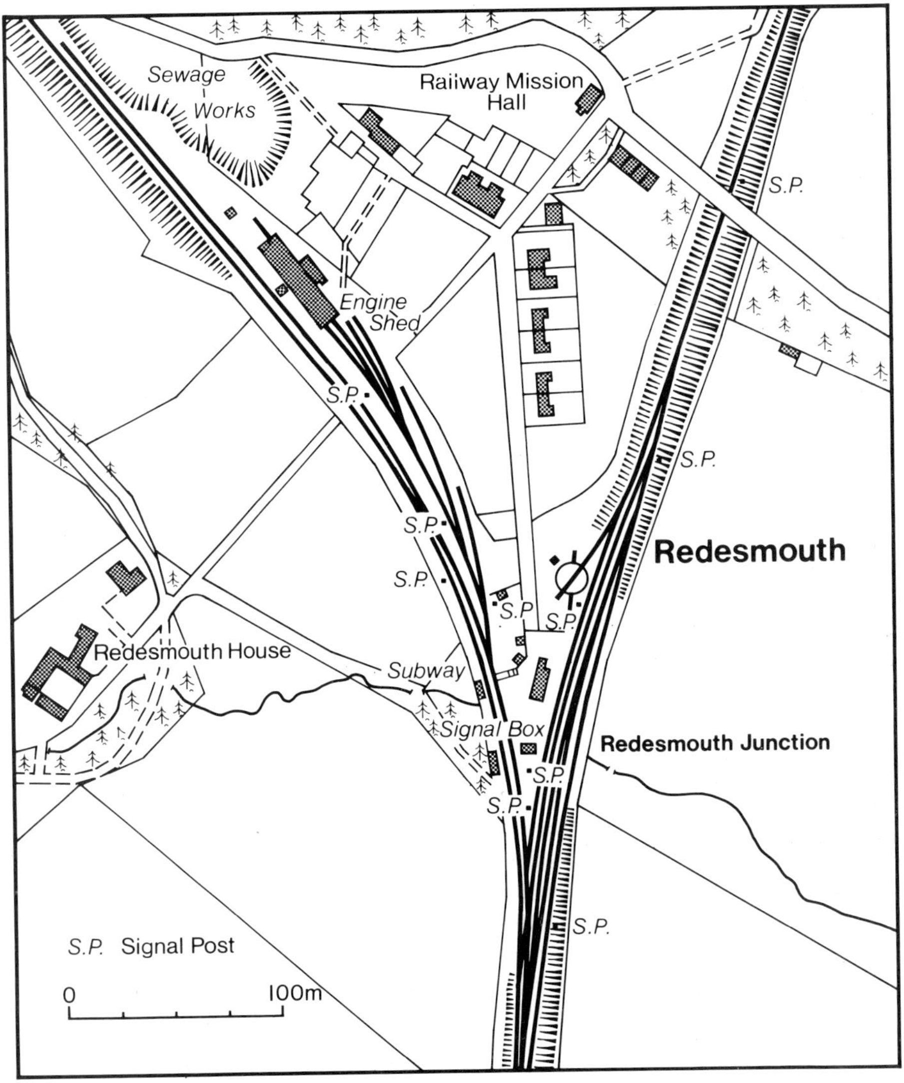

Reedsmouth Junction

All that the LNER got out of the development of this huge forest was some inward traffic of young trees for planting that were brought down from Aviemore and some additional passenger traffic from a newly-built Halt platform at Lewiefield, two miles south of Kielder. This platform with its diminutive waiting shelter was erected in 1936 to serve a Ministry of Labour training camp. The camp was established to provide work in forestry for unemployed miners and shipyard workers from Northumberland and Tyneside. They came to Kielder to erect fences and plant trees for the Forestry Commission. On the outbreak of war the same hutted camp was occupied by conscientious objectors who were directed to do forestry work. Later it was used to house displaced refugees from Europe.

Locomotives for passenger trains were provided by Hawick, Riccarton and Blaydon sheds except for the Saturday only train from Kielder to Hexham and back which was worked by a Reedsmouth engine, an ex-North British 0-6-0 LNER class J36. When this shed closed in 1952 the Saturday Kielder service was worked by a class V1 / V3 2-6-2- tank engine with coaching stock from Hexham. This involved 67 miles of empty train running for 67 miles of revenue earning running – a lesson in how to lose money quickly!

Enthusiasts of the Border Counties line have recorded 38 different classes of locomotive that have worked on it. Many of these were of "one-off" appearances on director's specials, stock trains and excursions. The basic class for many years, as I remember, was the ex-North British "Scott" class 4-4-0 of LNER class D30 and engines of similar D31 and D32 classes. The D30's were all named after characters in the novels of Sir Walter Scott and engines that are well remembered include "Caleb Balderstone", "Ellangowan", "Norna", "The Talisman", "Dugald Dalgetty" and "Wandering Willie". Two of these engines were stabled at Blaydon especially to work the Newcastle-Hawick trains. Eventually they were replaced by D49 3-cylinder "Hunt" class 4-4-0's and three engines of this class – "The Percy", "The Zetland" and "The Rufford" – were regular performers. Other classes tried out on passenger trains were ex-Great Northern 4-4-0's (LNER class D1) and ex-North Eastern 4-4-0's of class D20. Neither of these were successful because they were unable to maintain time and they threw out too many sparks in the vulnerable newly-planted forest. Towards the end of steam a number of strangers appeared on the line including B1 4-6-0's, K1 and K3 2-6-0's, Ivatt Class 4MT's and standard 2-6-0's of classes 2, 3 and 4.

Engines for goods trains gave much less variety. Those from Riccarton and Hawick were ex-North British 0-6-0's of classes J35 and J36 while Hexham and Blaydon sheds supplied ex-North Eastern 0-6-0's of class J21 with sometimes a Gresley J39 0-6-0.

Reference has already been made to the shed at Reedsmouth and its ex-North British 0-6-0's. These engines covered local services between Reedsmouth and Scot's Gap and also the Saturday service between Kielder and Hexham, but what else they did I don't know. Reedsmouth shed was also home for a time of North British No. 224, the infamous engine that was drowned in

Photo: Lens of Sutton

Humshaugh station when the goods yard was busy but after the demise of the colliery and coke ovens. The apparently collapsed building at the far end of the platform is a derrick for loading timber.

Photo: Ian S. Carr

A rare picture of a Class K3 2-6-0 on the Border Counties Line. No. 61879 from St. Margaret's shed Edinburgh enters Wark station with the 16-32 from Hawick on 14th August 1956.

Photo: Ian S. Carr

Class D49/2 4-4-0 No. 62771 "The Rufford" on a Hawick train at Reedsmouth on 9th June 1956. Two engines of this, or the "Shire" class were allocated to Blaydon shed for working the Border Counties Line. Reedsmouth shed is obviously closed!

Photo: Northumberland County Record Office

Class D1 4-4-0 (ex GNR) at Reedsmouth Junction on a train for Newcastle in July 1937. Engines of this class were brought North for trials on hilly routes, like the Border Counties Line. They were also used on the Stainmore Line between Darlington and Penrith.

Photo: C. J. B. Sanderson

Reedsmouth Junction with the solitary carriage from Bellingham to Scot's Gap waiting for the engine to run round. This picture gives a good view of the signal box and the very large water tank.

Photo: Ian S. Carr

Bellingham Show Day, 22nd September 1956. Class J39 0-6-0 No. 64814 stands at Reedsmouth with the 08-50 special train from Newcastle while Class J21 0-6-0 No. 65061 brings in empty coaching stock from Bellingham. Reedsmouth shed closed in 1952 hence the dilapidated state of the track.

Photo: Alex Appleton

Reedsmouth shed and coaling stage from the west side of the river.

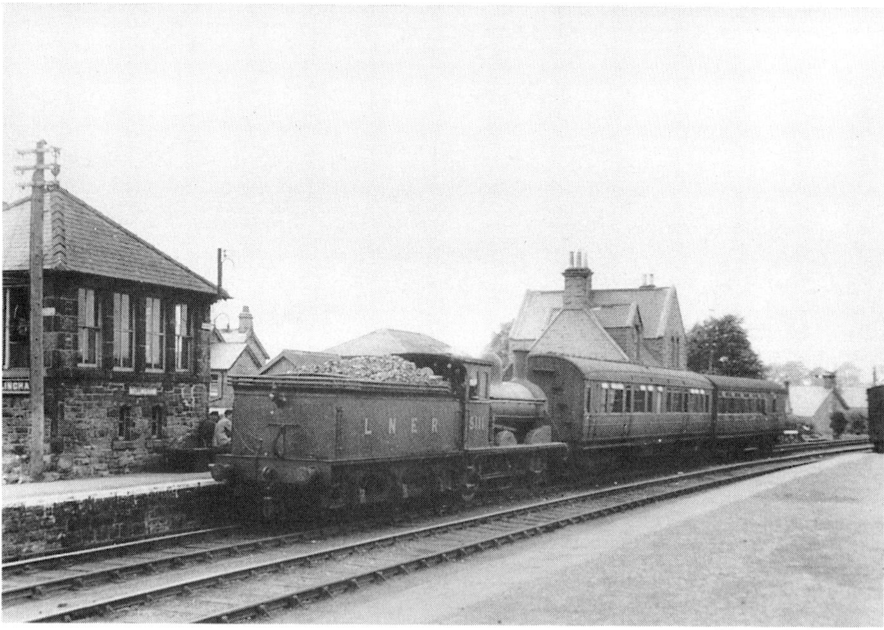

Photo: J. Dawson

Class J21 0-6-0 No. 5111 waits at Bellingham with the 4-05 p.m. to Reedsmouth and Scot's Gap.

the Tay Bridge disaster. Scotland's famous poet and tragedian, William McGonegall, described the event dramatically, thus:-

"So the train moved slowly along the Bridge of Tay,
Until it was about midway,
Then the central girders with a crash gave way,
And down went the train and passengers into the Tay.
The Storm Fiend did loudly bray
Because ninety lives had been taken away
On the last Sabbath day of 1879,
Which will be remembered for a very long time"

The engine was eventually recovered and reconditioned but superstitious Scottish drivers would have nothing to do with it so it was sent to Reedsmouth where the more realistic drivers of the North Tyne nicknamed it "The Dipper" and ran it for the rest of its days. Incidentally, although the location of this shed has always been spelled on the map as Redesmouth – it is at the mouth of the River Rede – the railway time-table has, for some unknown reason, always printed it as Reedsmouth.

In the days before Dr. Beeching wielded his axe, there were a number of stations in Great Britain that had no road access. They were quite isolated from any form of communication. Afon Wen, where the LMS line from Caenarfon joined the Great Western Cambrian Coast line, was one. Cairnie Junction on the Great North of Scotland line from Aberdeen to Keith was another. Corrour on the West Highland line had a road, of sorts, to outlying shooting lodges and crofts in the immediate vicinity but it did not connect with any other road. These places were stations or platforms with a very small resident population. Riccarton Junction, however, was a railway village with rows of terraced houses yet there was no road to it. It was in the middle of nowhere and entirely dependant on the railway for its existence and survival and when the railway went Riccarton went!

Some 32 miles from Carlisle and 65 from Edinburgh, Riccarton was situated 880 ft above sea level. The station had an island platform for main line trains and a bay at the south end for Border Counties line trains to Hexham and Newcastle. The platform housed the village shop and post office and everything that came into or out of Riccarton did so by train. There was no resident doctor in the village hence an engine was always kept in steam to seek medical assistance from either Newcastleton (9 miles) or Hawick (13 miles). There was no graveyard or cemetery at Riccarton so, since the railway brought all residents into the village, it ultimately took all of them out.

Remote indeed, yet, by catching the northbound train at 10-24 a.m. Riccarton folk could be in the centre of Edinburgh for lunch. Alternatively, the 9-13 a.m. train south made a connection at Carlisle that enabled passengers to be in Euston by 3-35 p.m. One would be hard pressed to do the same journey in the same time today.

One of the great advantages that trains give to walkers and ramblers is that they enable walks to be made that are neither circular nor involve returning along the same route on which one set out. I recall two very enjoyable days that were made possible by the Border Counties line.

They both involved staying at Falstone. The "Black Cock" in those days was an inn of character complete with privy in the garden! We took the 12-58 a.m. train to Riccarton then walked back over the hills to Saughtree. This walk enabled us to experience the wonderful silence of the Cheviot Hills at its best. At Saughtree we crossed the Liddle Water and climbed over Windy Knowe to the headwater of the North Tyne at Deadwater. We had no trouble in finding seats on the 5-01 p.m. train back to Falstone!

Another day we walked from Falstone to Lewisburn then along the Coal Road, that was built before the railway, to enable coal to be taken from Plashetts colliery to Liddlesdale and Newcastleton. The road first follows the course of the Lewis Burn and then its tributary, the Akenshawburn, which leads up to the border at Bloodybush, the site of a border battle. At Bloodybush, 1,312 ft, there is a magnificent stone pillar on which is inscribed the information that the English part of the road was the property of Sir J. E. Swinburne, Bart., of Capheaton and that the Scottish part belonged to William Oliver Rutherford. It also lists the rates of toll charged for each kind of merchandise and livestock transported over the road. The descent on the Scottish side led to Steel Road station where we caught the 4-14 p.m. train for the 12-minute journey to Riccarton. The ensuing wait of 22 minutes enabled us to sample some of the edible contents of the platform shop before leaving on the 4-48 p.m. train back to Falstone.

Alas, we have progressed since then and such pleasurable days are no longer possible.

The Border Counties was an interesting line on which to travel. It began in the rich productive farmland of the Tyne valley at Hexham and followed the North Tyne from its confluence with the South Tyne at the "Meeting of the Waters". Railway and river were never far apart until the source of the river at Deadwater in the sparsely populated hill sheep country that extends from Kielder to Saughtree and Riccarton. The railway traversed country that was attractive without being positively scenic. It also provided a valuable service to isolated communities in the valley. A former doctor at Bellingham has described how, by making a telephone call to Hexham station in the evening, the next morning's train would be stopped at remote farms to deliver urgent prescriptions put on at Bellingham. The train also stopped to pick up urgent hospital cases such as a mis-carriage or fracture when the patient, having been carried down the fell on a gate, would be lifted on to the train at the line side. There was also the tragic occasion when a Hawick driver, by the name of Young, broke the cardinal rule of footplatemen and climbed on to the tender to answer a call of nature. His head was knocked off by the overbridge at The Law, just north of Bellingham.

From time to time the LNER ran excursions from Newcastle to Hexham, Bellingham, Riccarton, Carlisle and back to Newcastle. I travelled on one in

1953, on a Sunday, ostensibly to view station gardens. The train consisted of six Gresley corridor carriages with a K1 2-6-0 at the front. The engine was turned twice during the trip, at Riccarton and Carlisle. It was a good day and the train was full. Why were such excursions not run more often? The only other special trains that I can recall were those on the third Saturday of September, the day of the Annual Show of the North Tyne Agricultural Society at Bellingham. This event called for extra carriages on trains from Hexham and special through trains from Bedlington and Morpeth and from Rothbury.

The line was single throughout with passing loops at Wall, Wark, Reedsmouth, Bellingham, Falstone and Kielder. In 1945 there were signal boxes at all of these stations with a gate box at Barrasford. Plashetts box was closed in 1925 and the one at Wall was destroyed by a fire in 1955. Signals remained North British until the end although some lower quadrant arms were replaced with upper quadrant ones. Posts were of lattice construction and the signal lamp was held in position behind the colour spectacle by a wire and pulley arrangement. When lamps required attention the lamp boy did not have to climb a ladder but merely lowered the lamp to ground level, did what was necessary and hauled it up again.

Signalling arrangements were as follows:

Miles from Hexham	*Location*	*Apparatus*
1.1	Border Counties Junction	Signal box, block post
1.8	Acomb Colliery siding	Ground frame
3.8	Wall station	Signal box, block post until destroyed by fire 1955
5.0	Chollerford (Humshaugh)	Ground framd
5.7	Cocklaw Quarry siding	Ground frame
6.2	Chollerton station	Ground frame
7.7	Barrasford station	Gate box, frame released by key
8.4	Barrasford Quarry siding	Ground frame
11.2	Wark station	Signal box, block post
15.2	Reedsmouth Junction	Signal box, block post
17.0	Bellingham station	Signal box, block post
20.5	Tarset station	
21.7	Thorneyburn station	Ground frame, signals
24.9	Falstone station	Signal box, block post
30.2	Plashetts station	Ground frame. Signal box closed 1925
31.7	Lewiefield Halt	
33.7	Kielder station	Signal box, block post
36.2	Deadwater station	Ground frame, signals
39.2	Saughtree station	Ground frame
42.0	Riccarton South	Signal box, ground frame

The only turn-table on the line was at Reedsmouth.

Closure of the Border Counties was anticipated in August 1948 when the same storms that breached the Wooler line caused severe damage to the Border Counties bridge over the Tyne west of Hexham. Again, as in the case of the Wooler line, the cost of replacing this bridge was beyond any immediate revenue that the line could earn. Temporary repairs undertaken in 1948 enabled the line to keep going until 1956 when the passenger service was withdrawn. Goods trains continued to use the line until 1958 when the Border Counties Junction to Reedsmouth and Bellingham to Riccarton sections were abandoned. Goods traffic continued to Bellingham until 1963 with trains being worked to and from Morpeth via Scot's Gap.

The Border Counties line came and went within a period of 100 years. Now it has all gone, some of it submerged by Kielder Water. Portions of the infrastructure remain. Kielder viaduct, built in 1867 and described as the finest example of a railway arch in Britain, is a listed building. But what is a railway viaduct without a railway?

When I go up the North Tyne valley now it is depressing to see where the railway once ran yet the memories of trains travelling on them remain as fresh as ever. Not least the sound of a "Scott" whistle echoing round the hills as it climbed the 1 in 100 from Deadwater to Saughtree. Once heard, never forgotten.

Table 163 HEXHAM and RICCARTON JUNCTION

Miles		Week Days only mrn A		mrn B		aft	aft	
	154 NEWCASTLE..... dep	5 55	..	1110	..	..	4 20	..
—	**Hexham**............ dep	6 51	..	1210	..	..	5 10	..
3½	Wall..................	6 59	..	1217	..	..	5 17	..
5	Humshaugh............	7 4	..	1222	..	..	5 22	..
6½	Chollerton............	7 9	..	1226	..	..	5 27	..
7½	Barrasford............	7 13	..	1230	..	..	5 31	..
11¼	Wark..................	7 22	..	1238	..	..	5 39	..
15½	Reedsmouth...... arr	7 30	..	1246	..	..	5 47	..
	Reedsmouth...... dep	7 46	..	1250	..	3 37	5 55	..
17	Bellingham (NorthTyne)	7 51	..	1254	..	3 41	5 59	..
20½	Tarset................	7 58	..	1 1	..	..	6 6	..
21½	Thorneyburn...........	8 2	..	1‡ 5	..	..	6 10	..
25½	Falstone..............	8 11	..	1‡11	..	..	6 18	..
30½	Plashetts.............	8 20	..	1‡20	..	..	6 27	..
32	Lewiefield Halt........	8 26	..	1‡25	..	..	6 32	..
33¾	Kielder...............	8 34	..	1‡32	..	..	6 39	..
36½	Deadwater.............	8 40	..	1‡38	..	..	..	..
42	Riccarton Junc..... arr	8 53	..	1‡50	..	..	6 54	..
55	167 HAWICK......... arr	1040	..	2F18	..	..	7 19	..
107¾	167 EDINBURGH (Wav) "	1226	..	4 49	..	..	9 58	..

Miles		Week Days only mrn B		mrn		aft		aft
	167 EDINBURGH (W.).dep	..	..	6 20	..	To Scotsgap (Table 164a)	..	2 25
	167 HAWICK......... "	6 5	..	8 41	..	..	..	4 25
—	Riccarton Junc..... dep	6 37	..	10 20	..		..	4 58
5½	Deadwater.............	Cc	..	10 30	..		..	5 8
8½	Kielder...............	6 53	..	10 35	..		..	5 14
10	Lewiefield Halt........	6 58	..	10 40	..		..	5 19
11½	Plashetts.............	Bb	..	10 44	..		..	5 23
16¾	Falstone..............	7 10	..	10 52	..		..	5 31
20½	Thorneyburn...........	7 17	..	..	..		..	..
21½	Tarset................	7 21	..	11 1	..		..	5 40
25	Bellingham (NorthTyne)	7 28	..	11 8	..	4 5	..	5 47
26½	Reedsmouth....... arr	7 32	..	11 12	..	4 9	..	5 51
	Reedsmouth....... dep	7 48	..	11 23	..	..	..	5 56
30¾	Wark..................	7 57	..	11 31	..	..	..	6 4
34½	Barrasford............	8 3	..	11 37	..	..	..	6 10
35½	Chollerton............	8 8	..	11 40	..	..	..	6 13
37	Humshaugh............	8 12	..	11 44	..	..	..	6 17
38½	Wall..................	8 17	..	11 48	..	..	..	6 21
42	Hexham........... arr	8 24	..	11 55	..	..	..	6 28
62¾	154 NEWCASTLE... arr	9 5	..	12 36	..	..	..	7 9

A Thro' Train between Newcastle and Riccarton Junction.
B Thro' Trains between Newcastle and Hawick
‡ 3 minutes later on Saturdays.
Bb Calls on Tues. and Sats. at 7 1 mrn when required to take up.
Cc Calls at 6 45 mrn. when required to take up on informing the Station Master at Riccarton Junc. before 50 aft the day previous to travel
F Arr. 2 24 aft. on Saturdays.
S Saturdays only.

Tables 164 & 164a MORPETH, SCOTSGAP, and REEDSMOUTH, and ROTHBURY

Miles		Week Days only mrn		aft S		aft		
	157 NEWCASTLE.... dep	9 30	..	1A15	..	5 §7	..	..
—	**Morpeth**.......... dep	10 15	..	2 15	..	5 50	..	..
5½	**Meldon**..............	10 27	..	2 28	..	6 3	..	..
7¾	**Angerton**............	10 34	..	2 34	..	6 9	..	..
9¼	**Middleton North**......	10 39	..	2 39	..	6 14	..	..
11¼	**Scotsgap** ◗..........	10 45	..	2 45	..	6 20	..	..
—	**Scotsgap**........ dep	10 51	..	..	..	6 27	..	..
14¾	**Knowesgate**..........	11 0	..	..	..	6 38	..	..
21¾	**Woodburn**............	11 13	..	..	..	6 52	..	..
25½	**Reedsmouth**..... arr	11 20	..	..	..	7 0	..	..
14½	Longwitton............	10 57	..	2 55	..	6 21	..	..
16½	Ewesley...............	11 2	..	3 0	..	6 38	..	..
17½	Fontburn Halt.........	11 7	..	3 4	..	6 43	..	..
22	Brinkburn.............	11 17	..	3 13	..	6F52	..	..
24¼	**Rothbury**......... arr	11 23	..	3 19	..	6 58	..	..

Miles		Week Days only mrn		mrn S		aft				
	Rothbury......... dep	7 50	..	1135	..	4 21	..	..	..	..
2¼	Brinkburn.............	7 55	..	1140	..	4 26	..	..	..	..
6¾	Fontburn Halt.........	8 6	..	1151	..	4 37	..	..	..	..
7¾	Ewesley...............	8 9	..	1154	..	4 40	..	..	..	..
9¾	Longwitton............	8 15	..	12 0	..	4 46	..	..	..	..
—	Mls **Reedsmouth**..dep	7 40	..	..	..	4 15	..	..	..	..
—	3¾ **Woodburn**..........	7 51	..	..	..	4 24	..	..	..	..
—	10¾ **Knowesgate**........	8 7	..	..	..	4 39	..	..	..	..
—	14 **Scotsgap** ◗... arr	8 16	..	..	..	4 48	..	..	..	..
13	**Scotsgap** ◗..........	8 24	..	12 7	..	4 53	..	..	..	..
15	Middleton North.......	8 29	..	1211	..	4 57	..	..	..	..
16½	**Angerton**............	8 34	..	1215	..	5 1	..	..	..	..
18¾	**Meldon**..............	8 41	..	1220	..	5 6	..	..	..	..
24¼	**Morpeth**........ arr	8 53	..	1230	..	5 16	..	..	..	..
40¼	157 NEWCASTLE ... arr	9 30	..	2 11	..	5 55	..	..	..	..

A Change at Manors.
◗ Station for Cambo (1 mile).
S Saturdays only.
F Calls to set down only.
§ Dep 5 0 aft on Sats.

From the L.N.E.R. (North Eastern Area) "Public Timetable", May-September 1945

The Wanney

When the Wansbeck Railway was promoted in 1857 the proposal was to build a railway that was completely independent of the North Eastern Railway, to link the Blyth & Tyne Railway at Morpeth with the Border Counties Railway at Reedsmouth. The North British Railway, ever eager to gain access to Newcastle, saw the possibility of having a route from Edinburgh to Newcastle via Riccarton, Reedsmouth, Morpeth and then by the Blyth & Tyne through Bedlington and Newsham into their New Bridge Street station even though it would be 13 miles longer than the main line from Edinburgh to Newcastle. Consequently the North British amalgamated with the Border Counties in 1860 and incorporated the Wansbeck Railway in 1863. Alas, their plans were thwarted when the North Eastern gained control of the Blyth & Tyne in 1874.

Although the Wanney was built to connect Morpeth with Reedsmouth it was never operated in this way. The completion in 1870 of the Northumberland Central Railway's line from Rothbury to Scot's Gap and its subsequent absorbtion by the North British in 1872 led to the line from Morpeth to Rothbury becoming the major one.

A curious feature of all this railway intrigue was that the junction at Reedsmouth faced Hexham so that trains from Edinburgh to Newcastle would have had to reverse at Reedsmouth. The Wansbeck Railway, as built, was most convenient for the running of through trains from Hexham to Morpeth but, with the exception of scenic excursions, this never happened.

The "Wanney" line was, in effect, the name given to the fourteen mile section between Scot's Gap and Reedsmouth and it was operated as a branch-line feeder service. Needless to say the "feeding" was at no more than subsistance level because the line ran through some very sparsely populated country. There were two intermediate stations at Knowesgate and Woodburn. At the time of the grouping in 1923 there were three daily passenger trains in each direction (four on Tuesdays) but by 1938 the 11-20 a.m. from Reedsmouth, which returned from Scot's Gap at 2-45 p.m., ran on Saturdays only. The latter, incidentally, was a through train to Bellingham.

Leaving Scot's Gap on the left-hand track (the right-hand one went to Rothbury) the two lines diverged after about a quarter of a mile. The Wanney line followed the shallow valley of Chesters Burn, a tributary of Hart Burn, climbing on to Kirkwhelpington Common. Knowesgate station, four miles from Scot's Gap, was sited where the railway crossed the Newcastle to Otterburn road (A696) but there was no village or community, merely Knowes House and some railway workers' cottages. Knowesgate was the station for Kirkwhelpington. At one time there was an active quarry to the east of the station with a connecting tramway and a number of sidings. There was also a passing loop and a signal box but this was closed in September 1924 and replaced by a ground frame with a "shut in" facility.

This rather cumbersome procedure involved releasing the ground frame by inserting the section token in a slide. When the train was "shut in" the loop, the points were restored to normal and the token was taken from the slide and inserted in an intermediate instrument located in the station booking office. Then, after consultation with the signalmen at Scot's Gap and Woodburn, token working was restored to enable another train to enter the single line. When the "shut in" train was ready to "escape" the token was retrieved from the intermediate instrument after consultation with the Scot's Gap and Woodburn signalmen and used to release the ground frame. After the train had "escaped", the ground frame was restored to normal and the token returned to the engine driver to continue his journey.

A mile west of Knowesgate the railway ran alongside the garden of Ray Demesne. This house was built and occupied by Sir Charles Algernon Parsons, the inventor of the steam turbine, until his death in 1931. Sir Charles had a private halt station, known as Ray platform, built for his own use and it is said that he had his own private railway carriage to take him to Newcastle each day. After his death the platform continued to be used by estate staff and their families and there are people alive today who travelled to school in Morpeth every day by train from Ray platform. As recently as May 1949 the North Eastern Region General Notices carried the instruction that passengers alighting or joining trains at Ray platform must pay the fare to or from the station beyond. Alas, the house became infected with wet rot and had to be demolished. The platform degenerated by natural causes and was no longer in use when the line closed in 1952.

Leaving Ray platform, the railway climbed around the south side of Ray Fell to the 850 ft summit of the line at Risey Burn with Great Wanney Crag to the south. Six and a half miles separated Knowesgate from the next station, Woodburn, and a very lonely stretch of line it was with no sign of any human habitation for most of the way.

Woodburn station was situated where the railway was crossed by the road from Corbridge to Carter Bar (A68). It was a rail-head for Otterburn and Redesdale and had a number of sidings for coal traffic and for the large quantities of stores and equipment that were brought in by train for the army ranges and training camps in the Otterburn area. There was also a railway connection into the adjacent Woodburn Quarries. Woodburn station was fully signalled and had a signal box and a passing loop. It remained a block post until the line was closed.

When, in 1893, the Newcastle and Gateshead Water Company embarked on the construction of a new catchment reservoir on the headwaters of the River Rede at Catcleugh, Woodburn became the southern terminus of a contractor's railway. The line was of 3 ft gauge and about 15 miles long. It was built to carry clay, stone and cast iron pipes to the site of the dam. At the peak of operations, eight locomotives were in use. This narrow gauge railway was operated for about ten years but there are very few traces of it left today.

About a mile and a half to the sout-west of Woodburn station are the remains of the industrial complex of Ridsdale. This had a fascinating history and the reader's tolerance is requested for the following digression.

Way back in the 1830s, long before the railway was built, the successful replacement of charcoal by coke in the iron making process led to the exploitation of coal and ironstone deposits at Ridsdale. Coke ovens (the beehive type) were erected to convert coal to coke and blast furnaces constructed to convert ironstone to iron. By 1845 there were 70 coke ovens and three blast furnaces with a wagonway linking the coal mines and quarries to the coke ovens and furnaces. The air blast was provided by a 120 h.p. beam engine and a 70 h.p. water wheel. The engine house for the beam engine can still be seen as a gaunt ruin on the west side of the A68 road as it descends from the south into the Rede valley.

By 1848 several puddling furnaces had been built and iron production reached 100 tons a week but, without a connecting railway, the movement of this amount of iron was very difficult. Nevertheless, the superior quality of Ridsdale pig iron was such that Hawks Crawshay, the Gateshead ironfounders, included 20% of it in the construction from 1846-49 of the High Level Bridge across the Tyne at Newcastle. The mind boggles at the thought of how they ever conveyed such quantities of heavy iron by horse and cart the forty or so miles from Ridsdale to Gateshead on rough roads with steep gradients where the average speed could have been little more than 2 m.p.h. Transport costs had become so high by 1852 that the entire undertaking was closed and moth-balled.

In 1864 Ridsdale was taken over by the Elswick Iron Company as a source of iron ore but iron making was not resumed and at least two of the blast furnaces were transferred to Elswick.

The completion of the Wanney line in 1865 brought a second era of prosperity to Ridsdale. Roasting kilns were erected in the area around Broomhope Burn and mineral workings at this site gradually took over from

Photo: C. J. B. Sanderson

Plashetts station after the colliery was abandoned and the signal box closed. The line is now single track but the water tank remains.

Photo: Northumberland County Record Office

Plashetts Colliery incline with the station in the middle distance. When the Border Counties Line was built there were great hopes that Plashetts would provide substantial mineral traffic to the Tweed valley towns and their mills. Alas, it never materialised.

Photo: Northumberland County Record Office

Lewiefield Halt erected in 1936 to serve the Ministry of Labour training camp (behind the telegraph pole). This camp was first used by unemployed miners and shipyard workers, then by conscientious objectors and finally by displaced workers from Europe.

Photo: Ian S. Carr

An immaculate Class VI 2-6-2T No. 67639 runs round its one coach train at Kielder Forest on 28th July 1956.

Photo: Allan W. Stobbs

Riccarton Junction from the South-east. The line in the distance is climbing to Whitrope Summit, two miles away.

Photo: Ian S. Carr

Two Class D30 4-4-0's, No. 62422 "Caleb Balderstone" and No. 62423 "Dugald Dalgetty" simmer at Hawick on 27th July 1956. Both engines were regular performers on the Border Counties Line.

Photo: R. W. Lynn Collection

Naval Gun train en route to Broomhope proofing range. This could be a 16 inch gun for H.M.S. "Nelson" or H.M.S. "Rodney".

Photo: Northumberland County Record Office

Class J21 0-6-0 (ex NER) at Woodburn with an excursion to a wet Bellingham Show.

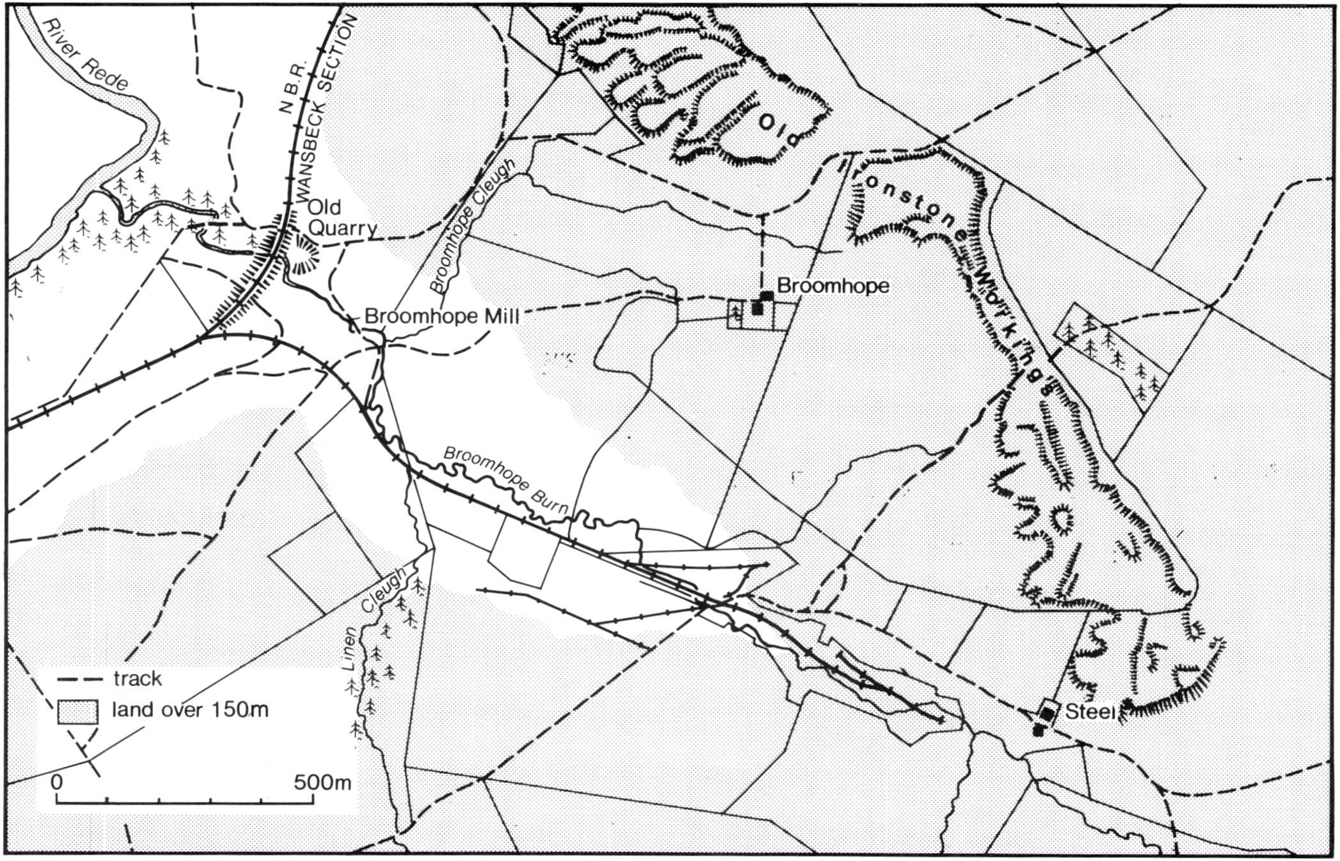

Broomhope siding

those at Ridsdale. Local coal was used to roast the ironstone to produce sinter which was sent by train along the spur railway in the Broomhope valley to its junction with the Wanney line. There was another halt platform at this junction, latterly known as Vickers platform.

In the six months from August 1867 to January 1868, 20,603 tons of sintered ore (73% of all traffic to Newcastle) were carried from Broomhope for smelting at Armstrong's foundry at Elswick. Output peaked in the 1870s and in 1876 new sinter kilns were built at the head of Broomhope branch. However, competition from imported high grade iron ores from Spain, Sweden and South America became so great that all workings at Ridsdale and Broomhope ceased in 1879.

The Ridsdale workings were owned and developed by W. G. Armstrong & Co. and when sinter production came to an end the site was put to an alternative use as a proof range for the testing of military and naval guns built at their Elswick works. The range has continued in the ownership of successive companies, first Armstrong Whitworth, then Vickers Armstrong and now Vickers Defence Systems. The peak of activity was during the first World War when over 10,000 guns were proved and more than 60,000 rounds were fired. In the 1920s the mighty 16″ guns for the battleships H.M.S. "Nelson" and H.M.S. "Rodney" were proved at Ridsdale. They were taken there on special trains of 140 ton "Gunset" bogie wagons with a brake van and a J21 0-6-0 at each end. The spur line, with points facing towards Reedsmouth, remained in being, though little used, until the Wanney line closed.

A mile beyond Broomhope spur the Wanney line joined the Border Counties line at Reedsmouth Junction.

Traffic over the "Wanney" was never very heavy. In the early days of the LNER, when the North British had been absorbed by the new company, there were three trains a day in each direction. They connected at Scot's Gap with trains between Rothbury and Morpeth. The first train of the day left Reedsmouth at 8-00 a.m. and arrived at Scot's Gap at 8-35 a.m. to connect at 8-44 a.m. with the train from Rothbury which included a through carriage to Newcastle. The Reedsmouth train then "sat" at "The Gap" until 10-22 a.m. when, after the departure of the first train to Rothbury at 10-17 a.m., it left for Reedsmouth arriving there at 10-55a.m.

The second train left Reedsmouth at 11-20 a.m. and arrived at Scot's Gap at 11-53 a.m. to connect with the late morning train from Rothbury. This left Scot's Gap at 12-05 p.m. The return train awaited the arrival of the one from Morpeth at 2-45 p.m. and left nine minutes later to arrive at Reedsmouth at 3-27 p.m. where, after reversal, it proceeded to Bellingham, arriving there at 3-41 p.m. It left Bellingham at 4-05 p.m. and, after another reversal at Bellingham, it made its final journey of the day to Scot's Gap at 4-15 p.m. where it connected with the 4-29 p.m. from Rothbury which called at Scot's Gap at 5-02 p.m. The return to Reedsmouth was at 6-32 p.m. (except on Tuesdays when it was 6-45 p.m.) after the departure of the evening train from Morpeth to Rothbury.

Connections at Reedsmouth were not as good as those at Scot's Gap. The morning departure at 8-00 a.m. did connect with trains from both Riccarton and Hexham which passed each other at Reedsmouth between 7-39 a.m. and 7-46 a.m. The 10-55 a.m. arrival from Scot's Gap and subsequent return at 11-20 a.m. gave a reasonable connection with the Riccarton to Hexham train which called at Reedsmouth at 11-04 a.m. Other connections were poor, except on Tuesdays when there was an additional train from Scot's Gap at 5-08 p.m. This arrived at Reedsmouth at 5-41 p.m. to connect with trains for both Riccarton and Hexham which passed at Reedsmouth at about 5-45 p.m. These two trains gave good connections with the Tuesdays only 6-05 p.m. to Scot's Gap but this arrived after the departure of the 6-27 p.m. to Rothbury. Consequently, while it was possible to have a fairly convenient journey from either Riccarton or Hexham to Morpeth, it was very difficult to travel between Riccarton or Hexham and Rothbury.

By the time that I used the line in the early 1940s the service had been reduced to two trains a day. The morning train consisting of a solitary ex-North British brake composite coach was hauled by an ex-North British 0-6-0 (LNER class J36). It left Reedsmouth at 7-47 a.m. and arrived at Scot's Gap at 8-32 a.m. to connect with the morning train from Rothbury to Morpeth. The engine was then turned on the turn-table and simmered in the turn-table siding until the arrival of the first train from Morpeth. After this had departed for Rothbury, the J36 backed its coach into Scot's Gap's single platform and, eventually, departed for Reedsmouth at 10-00 a.m. calling at Knowesgate and Woodburn en route. In the afternoon the J36 took its coach one and three-quarter miles up the North Tyne to Bellingham at 3-57 p.m. and brought it back again at 4-05 p.m. to form the 4-14 p.m. from Reedsmouth to Scot's Gap. There, it connected with the afternoon train from Rothbury to Morpeth. The final run of the day was the 6-22 p.m. from Scot's Gap to Reedsmouth.

On Mondays, however, things were different. There was a goods train! This left Reedsmouth at 11-30 a.m. and returned from Scot's Gap at 1-15 p.m.

I find it quite incredible that the nationalised British Rail should be subject to so much criticism over low labour productivity when the privately owned LNER employed six men (2 drivers, 2 firemen and 2 guards) to move one railway carriage 59 miles in a day. People today have no idea as to the inefficient way in which private railway companies were operated.

During the war and as long as National Service lasted, the War Department provided a lot of additional traffic to Woodburn. This involved troop trains taking men to, or from the Otterburn and Redesdale ranges and training areas, and special goods trains conveying wheeled and tracked vehicles, field guns, ammunition and general stores. There was also an increase in the amount of traffic using Broomhope siding.

Reference has already been made to signalling arrangements at Knowesgate and Woodburn and these are included in the following line summary:-

Miles from Scott's Gap	*Location*	*Apparatus*
3.6	Knowesgate Quarry	Ground frame
4.2	Knowesgate station	Intermediate instrument, ground frame and signals
5.6	Ray platform	
10.7	Woodburn station	Signal box, block post
12.5	Broomhope siding and Vickers platform	Ground frame
13.0	Reedsmouth Junction	Signal box, block post

The only other use to which the line was put was as part of the scenic circular tours that the LNER ran in the 1930s. The route varied from time to time but usually followed the course from Newcastle to Morpeth via the Blyth & Tyne railway (so as to avoid a reversal at Morpeth), then to Scot's Gap, Reedsmouth, Hexham and back to Newcastle. The early excursions were composed of teak open saloon coaches but later green and cream liveried tourist stock was introduced. Passengers paid the fantastically cheap fare of three shillings and sixpence (17½p) for the 82-mile journey!

After the closure of the Border Counties line in Speptember 1958, the Wanney became the only railway access to Bellingham and the North Tyne valley. Then, in 1963 when the Rothbury line was closed, the Wanney was cut back to Woodburn and operated as one section from Morpeth station signal box (Wansbeck box at Morpeth was closed at the same time) to Woodburn. Military traffic for Otterburn kept the line open for a further three years as the last remaining remnant of the North British Railway in England. It was worked by J27 0-6-0's from Blyth shed and by new class 25 diesel-electrics.

Finally, on Sunday, 2nd October 1966 the "Wansbeck Piper" played a requiem pibroch. It was organised by Gosforth Round Table as the last steam-worked passenger train from Newcastle to Morpeth and Woodburn. The eleven maroon corridor coaches carried 640 passengers from all over the country and were hauled by two "foreign" engines, Ivatt class 4MT 2-6-0's No's 43000 and 43063 coupled tender to tender so that the leading loco was always running chimney first.

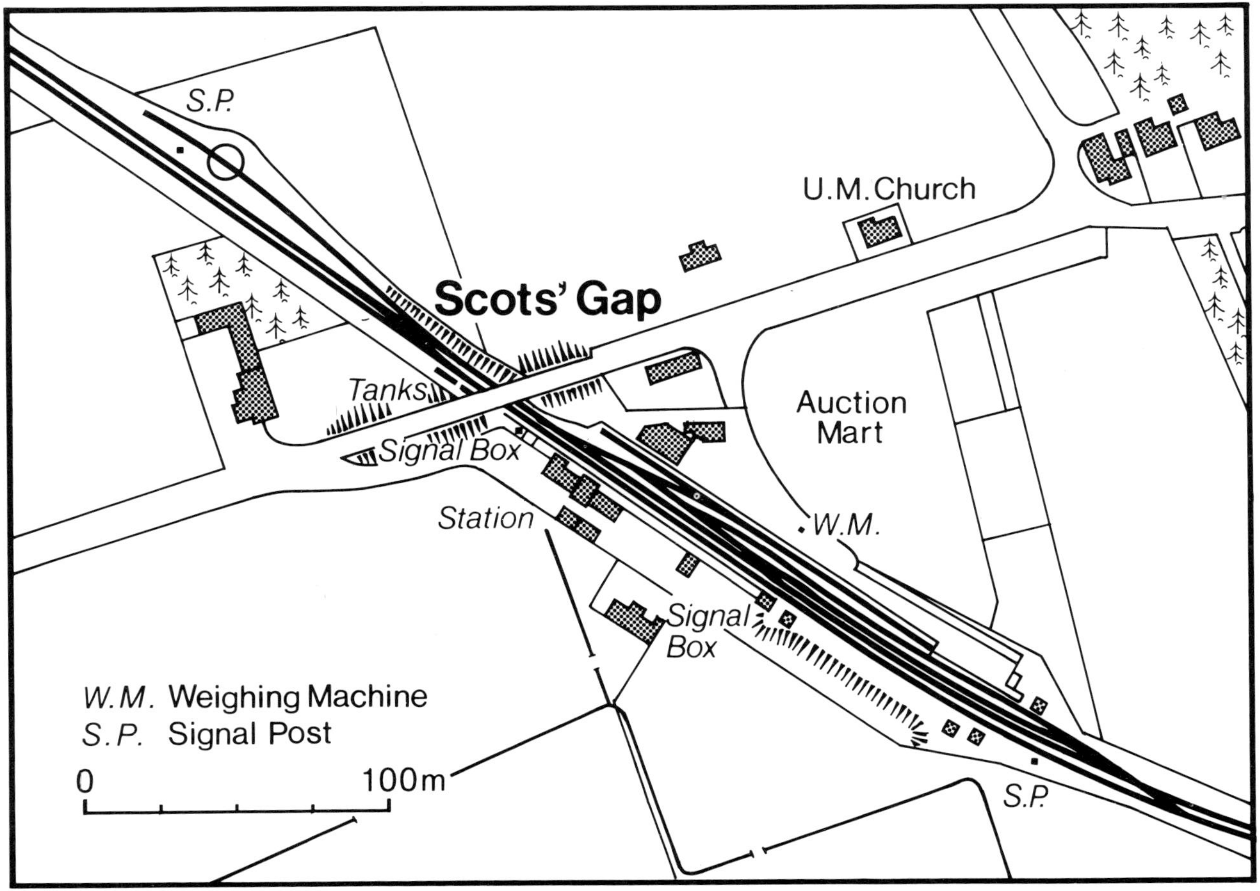

Scot's Gap station

The Rothbury Line

The railway from Morpeth to Rothbury was built in two stages. The first was the Wansbeck Railway which was authorised to construct a line from Morpeth to Reedsmouth where it would link up with the Border Counties Railway which opened in February 1861. The Morpeth to Scot's Gap section of the Wansbeck Railway was opened in July 1862. It was amalgamated with the North British Railway the following July.

The second stage from Scot's Gap to Rothbury was not completed until November 1870 when the 13-mile line was the residue of the grandiose Northumberland Central Railway promoted to run from Coldstream to Scot's Gap via Wooler, Wooperton, Alnham and Rothbury. Lack of financial support caused the section from Coldstream to Rothbury to be abandoned and even the impoverished length from Rothbury to Scot's Gap lasted only until July 1872 before it too was amalgamated with the North British.

Although built by two separate companies, the line from Rothbury to Morpeth, when once in the hands of the North British, was regarded as the "main" line with the remainder of the erstwhile Wansbeck Railway to Reedsmouth being operated as a branch. This practice continued into LNER days when I got to know the line.

Leaving Morpeth on a gradient of 1 in 90, the line eased to a gentle rise up the broad valley of the River Wansbeck. The very names of the stations bring back the nostalgia of journeys made on it. There was Meldon, a block post and passing place until 1940; Angerton, also with a passing loop and level crossing; Middleton North (to distinguish it from all the other Middletons) in an exposed position on the north side of the valley; and Scot's Gap with its signal cabin, turn-table, sidings and cattle market.

Scot's gap was the junction where the Rothbury line and the Wanney line went their separate ways. Although what appeared to be double track left the west end of Scot's Gap station it was, in fact, two single lines which ran parallel for the first quarter of a mile, then diverged. The right-hand track veered north and

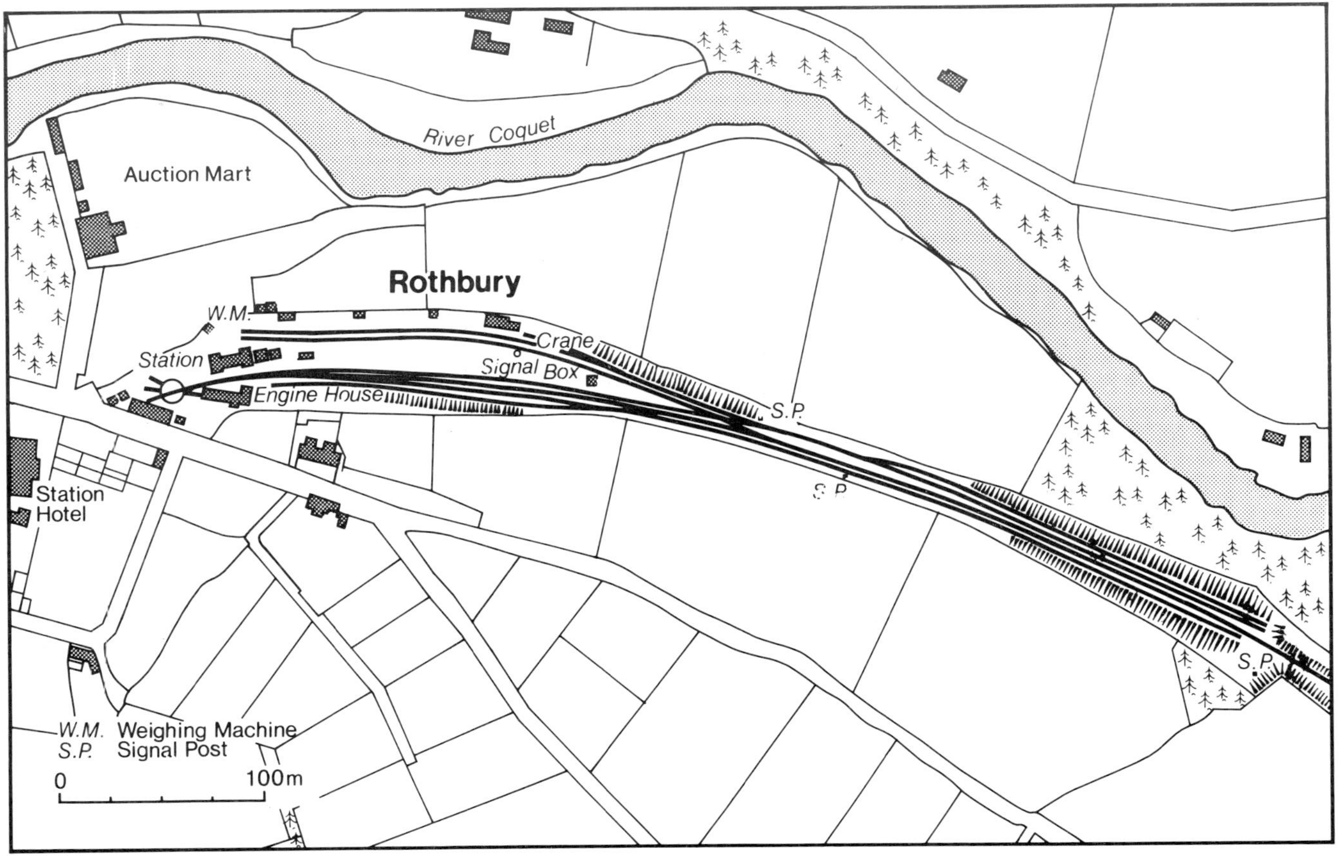

Rothbury station

immediately began to climb at 1 in 75 for the three miles to Longwitton station where the railway passed under the B6342 Hexham to Rothbury road. Like all the other stations on the line, Longwitton had a single platform with a booking office, waiting room and a small shed – in this case the body of an old carriage – for the storage of small goods consignments. There was also an open-air gentlemen's toilet! – now at Beamish. At the time that I knew Longwitton, in the 1940s, there was one siding but in the early 1900s it was a much busier place.

Shortly after leaving the station the railway reached a height of 694 ft, the summit of the line, before falling gently into the upper reaches of the Font valley and Ewesley station. This was very similar to Longwitton and, until April 1941, it had signals and was a block post but not a passing place. This arrangement broke-up the 13-mile section from Scot's Gap to Rothbury and so enabled two trains to travel in the same direction at the same time.

After leaving Ewesley the line curved to the right to cross the River Font by a magnificent stone viaduct, the grandest construction work on the line. It had no less than 12 arches each of 30 ft and at the time of writing (1992) is still standing in all its isolated glory.

Fontburn Halt was situated a short distance beyond the viaduct and consisted of a short platform on which there was a waiting shelter that was only slightly bigger than a watchman's hut.

Thereafter the line climbed again past quarries and defunct coal mines to a second summit at about 650 ft before descending on varying gradients, one as steep as 1 in 60, to Brinkburn. The buildings here were very similar to those at Longwitton and Ewesley and there was also a single siding trailing in from the Rothbury end of the platform.

The line continued to fall to cross Witton Burn, followed by a short ascent at 1 in 80 before the final descent of 1 in 75 to the end of the line.

Rothbury station had all the facilities expected at a rural branch line terminus. The single platform was longer than usual in order to accommodate the eight and nine coach trains that were run in the days when horse racing took place on the river haugh to the west of the town. The terminal end of the platform was covered by a canopy to protect passengers using the two and three coach trains that were normally run. The platform road terminated in a 42 ft turn-table that gave access to both the run-round loop and single road engine shed with its adjacent water tank. The turn-table was locked in position by a lever in the signal box. There were three sidings in the goods yard with ample provision for coal and van traffic, although the goods shed was smaller than might have been expected. Unlike North Eastern stations, there were no coal cells whereby coal could be dropped from hopper wagons in an elevated position. At Rothbury, coal had to be shovelled out! Rothbury has an important livestock mart and the goods yard was well equipped to deal with large numbers of cattle and sheep in the days when they were transported by train. Three long sidings at the east end of the station were usually packed with empty cattle wagons at the time of livestock sales. The station was fully signalled and the signal box was a block post, latterly for the 13-mile section to Scot's Gap.

Photo: Ian S. Carr

Class 4MT 2-6-0 No. 43129 pauses at Woodburn on 9th November 1963 with the RCTS/SLS "Wansbeck Wanderer" returning from Bellingham to Scot's Gap.

Photo: R. W. Lynn Collection

Scot's Gap - "The parting of the rails" - left to Reedsmouth, right to Rothbury.

Photo: R. W. Lynn Collection

Class J36 0-6-0 No. 9779 leaving Scot's Gap with its train of one ex-North British brake composite carriage for Reedsmouth.

Photo: Ian S. Carr

Class J21 0-6-0 No. 65103 approaches Scot's Gap with a goods train from Hexham via Reedsmouth on 8th August 1958. The line from Border Counties Junction to Reedsmouth closed on 1st September 1958.

Photo: Ian S. Carr

Class J25 0-6-0 No. 65727 at Scot's Gap on 8th August 1958 with the Morpeth to Rothbury goods. Note the cattle wagons having probably brought in Irish cattle for Scot's Gap mart.

Photo: A. J. Wickens

Longwitton station of fond memory. Our campsite was by the trees at the left of the picture. The new concrete-sleepered track is obviously the harbinger of closure!

Photo: Stuart Sellar

Ivatt Class 2MT 2-6-0 No. 46474 on a Rothbury goods at the summit of the line, east of Longwitton station.

Photo: Alex Appleton

Class J21 0-6-0 No. 877 at Rothbury on 5th August 1939 with the 11-30 a.m. train to Morpeth. With at least five carriages and a van, this was a heavy train for one engine.

Signalling arrangements on the Rothbury line were as follows:-

Miles from Morpeth	*Location*	*Apparatus*
5.3	Meldon station	Intermediate instrument, signals and ground frame
7.5	Angerton station	Gate box, signals
9.0	Middleton North station	Ground frame
11.0	Scot's Gap	Signal box, block post
14.3	Longwitton station	Ground frame
16.3	Ewesley station	Ground frame, signals, block post
17.5	Fontburn Halt	Ground frame
19.3	Forestburn siding	Ground frame
21.8	Brinkburn station	Ground frame
24.0	Rothbury station	Signal box, block post

Rothbury is a small country town with a population of about 1,600. It is situated in a well-farmed valley that produces good crops and livestock. The arrival of the railway caused the establishment of a livestock auction mart dealing with large numbers of cattle and sheep from the surrounding area and the North British and LNER had the responsibility of transporting these animals to all parts of the country. That was in the days when the welfare of livestock was considered to be important. All livestock had to be watered every four hours when they were moved by train.

Rothbury also developed as a popular holiday centre for Tynesiders and some of their wealthier businessmen, like Lord Armstrong, lived there and travelled daily to Newcastle by train. Indeed, even though the North British and North Eastern were rival companies, there was a "through" carriage from Rothbury to Newcastle in the morning and one back in the evening. Whether this was provided at the instigation of Lord Armstrong or speculatively for his benefit is not known but the practice continued after the formation of the LNER and only ceased with the outbreak of war in September 1939.

The LNER "through" working involved a brake composite coach carrying roof destination boards with the legend NEWCASTLE & ROTHBURY and ran morning and evening in both directions. The 8-05 a.m. train from Rothbury arrived at Morpeth at 9-08 a.m. where the through carriage was attached to the rear of the 8-35 a.m. train from Alnwick and arrived at Newcastle at 9-45 a.m. This carriage subsequently returned to Morpeth on the 1-33 p.m. from Newcastle to Berwick. It then left Morpeth at 2-15 p.m. and arrived at Rothbury at 3-19 p.m.

In the summer of 1938 there were four trains a day between Morpeth and Rothbury. From Mondays to Fridays the last train from Rothbury departed at 7-22 p.m. and ran express to Scot's Gap, calling at only Middleton North, before arriving at Morpeth at 8-05 p.m. Here the through carriage was attached to the 6-36 p.m. from Berwick to Newcastle and arrived at the Central station at

8-35 p.m. On Saturdays the last train left Rothbury at 8-28 p.m. and called at all stations to arrive at Morpeth at 9-24 p.m. The through carriage was attached to the 8-00 p.m. from Berwick and arrived at Newcaslte at 10-13 p.m.

In the reverse direction the evening carriage to Rothbury was attached to the 5-07 p.m. from Newcastle to Alnwick. Thirteen minutes were allowed at Morpeth to shunt it on to the Rothbury train which departed at 5-45 p.m. to arrive at Rothbury at 6-51 p.m. It is said that these through workings were not so much for the benefit and convenience of passengers as to avoid handling parcels and milk churns across platforms, and the busy main line, at Morpeth.

Although the railway went through countryside that was thinly populated, there were a number of quarries and small coal mines between Scot's Gap and Rothbury that contributed useful traffic in the early days. Longwitton station, situated at Rothley Shiel and about three miles from Longwitton village, was an interesting and busy place. To the west of the station, on the Scot's Gap side, a narrow gauge tramway brought lime to a siding from Rothley lime kilns while at the east end another tramway brought coal 1½ miles across the moor from Longwitton colliery to exchange sidings adjacent to the Rothbury line. A feature of Rothley is the two lakes, one on either side of the Hexham to Rothbury road. The lake on the east side was created by the erection of the embankment to carry the Longwitton colliery tramway over Rothley Gill and the source of Ewesley Burn. Ironically, in 1981-82, the abandoned coal deposits at Longwitton colliery were extracted by open-cast working and taken away by road transport, the railway having been itself abandoned while the coal was still underground.

Longwitton has many personal memories because, in 1942 and the following years, we had a number of Scout camps there. The campsite was on the south side of the railway with tents pitched in the excavated area where the exchange sidings for Longwitton colliery tramway had once been. Looking back, it was quite an operation getting 35 boys and all their equipment – tents, dixies, cooking utensils, sleeping bags, blankets and personal kit – from North Shields to Longwitton by train. Everything was assembled on a Friday evening and locked in an un-used waiting room at North Shields station. On Saturday morning it was loaded on to barrows for quick transfer to the luggage compartment of the electric train at about 11-45 a.m. In those days, stopping trains for Morpeth and Alnwick departed from Manors North, not Newcastle Central. Hence, on arrival at Manors East, all our equipment had to be unloaded and carried over the footbridge to Manors North bay platform and loaded into the luggage van of the Alnwick train. This departed at 1-12 p.m. and was routed by Jesmond and Benton where it joined the main line by a west to north curve that had been installed at the beginning of the war.

At Morpeth it was a case of "all hands on deck" to transfer our gear from the van of the Alnwick train across the platform and into the van of the waiting Rothbury train. This departed at 2-15 p.m. and there wasn't much space left in the luggage compartment of the ex-North British brake composite carriage when it left Morpeth! We arrived at Longwitton at about three o'clock and, after unloading everything out of the van, it all had to be carried along the lineside and lifted over the fence to the site of the camp. Eventually tents were pitched, a fireplace made, wood collected and some semblance of order achieved. It was our aim to have all this done by the time that the train returned from Rothbury at ten minutes to five – just to prove that we could do it. One year, it was so wet that we spent the first two hours in the station's diminutive waiting room until the rain eased off. We managed to light the fire and, with thirty or so people packed in, the atmosphere soon became distinctly humid!

Being so near the railway, we tended to use the trains as a timepiece. Breakfast was supposed to be ready by the time the morning train from Rothbury came by at 8-15 a.m., always hauled by Rothbury's own G5 0-4-4 tank No. 1918. The evening meal coincided with the passing of the train to Rothbury at 5-25 p.m. when we made suitable gestures and grimaces to the onlooking passengers.

The Rothbury line may not have been as busy as the main line but we saw everything that there was to see including light engine running and heavy military stores trains and extra livestock trains that ran in the evening.

But enough of Longwitton in the 1940s. In earlier days, lime was loaded at Fontburn from Ewesley quarry, this being three-quarters of a mile away and linked by a tramway. Coal was also loaded at Fontburn from Whitehouse colliery connected by another tramway; at Forestburn Gate from Forest Burn drift mine; and to another siding at The Lee from Lee coliery half a mile away.

When, in 1901, the Waterworks Department of Tynemouth Corporation built a new catchment reservoir on the River Font to the west of the railway, a special siding was installed to handle construction materials and transfer them to a contractor's railway. This siding was controlled by a ground frame.

Regretably, all these activities were in the past and well before I became aquainted with the line. The exception was Greenleighton Lime Works at Rothley where calcium limestone continued to be burned in spite of the black-out. The occasional wagonload of quicklime was loaded at Longwitton siding, sheeted, and dispatched to Bathgate. Even then, with petrol rationed, most of the output of the works was delivered by road. There was also timber from woods in the Font valley and this was loaded at Ewesley.

The passenger service in 1942 consisted of two daily trains in each direction, Mondays to Saturdays, with an additional mid-day service on Saturdays. The morning train left Rothbury at 7-50 a.m. and arrived at Scot's Gap at 8-24 a.m. where it picked up any passengers that may have arrived on the 7-40 a.m. from Reedsmouth. Arrival at Morpeth was at 8-53 a.m. – one hour and three minutes for 24½ miles. The return train left Morpeth at 10-08 a.m. and arrived at Scot's Gap at 10-38 a.m. where it was greeted by a stentorian female porter who announced in a very broad Northumbrian accent, "Scot's Gap, change for Knowesgate, Woodburn and Reedsmouth!".

This morning train from Morpeth also had a very important ancillary function. It delivered the newspapers. From Longwitton to Rothbury each isolated lineside cottage would receive a whistle from the engine as the paper was thrown from the cab into the garden. Arrival at Rothbury was at 11-16 a.m. and the engine was turned on the turntable even though it was a tank engine. I never saw it running bunker first on a train, in either direction. This operation completed, it was left to simmer until required for the afternoon train at 4-25 p.m. The evening train from Morpeth departed at 5-45 p.m., arrived at Scot's Gap at 6-15 p.m. with a connection to Reedsmouth at 6-22 p.m., and finally arrived at Rothbury at 6-52 p.m. – the end of the day. It always strikes me as being rather silly that a line as rural and relaxed as this one was should have train times as precise as 8, 22, and 52 minutes past the hour. Would it really have mattered and would the service have been jeopardised if trains had departed at 10, 20 and 50 minutes past the hour, or, to put it in the popular vernacular, ten past, twenty past and ten to the hour?

Rothbury had a shed for two engines. One for the passenger train the other for the goods. As from November 1940 the passenger engine was a Wilson-Worsdell 0-4-4 tank, LNER class G5, number 1918. This engine operated the service for the remaining LNER days. It remained at Rothbury as number 67296 after nationalisation until the shed closed in 1952. It was very much Rothbury's engine.

The goods train left Rothbury after the morning passenger had arrived at Scot's Gap. When Ewesley was still a block post, the goods could leave Rothbury when the passenger had cleared Ewesley but, as previously mentioned, Ewesley ceased to be a block post in April 1941. The goods would do some shunting at Scot's Gap and then set off for Morpeth, possibly allowing the morning passenger from Morpeth to pass it by being "shut-in" at Meldon. The afternoon return trip was made in time for the goods to arrive at Rothbury before the departure of the 4-24 p.m. passenger to Morpeth. Between Scot's Gap and Rothbury, all sidings were laid with trailing points from Scot's Gap so that all picking-up and setting-down had to be done on the afternoon run to Rothbury.

On Saturdays, operations were a little more complicated. There was an additional passenger train from Rothbury at 11-30 a.m. which called at Scot's Gap at 12-02 p.m. and arrived at Morpeth at 2-25 p.m. Consequently, unless the return goods cleared Morpeth by about 11-40 a.m. it could not pass the mid-day passenger at Meldon. Under these circumstances the goods had to remain at Morpeth until after the arrival of the passenger. If, in addition, goods traffic was heavy and involved a lot of station work, as was usually the case during the war, then the goods could be delayed at Scot's Gap until after the arrival of the passenger at Rothbury. If this happened, then the goods had about 65 minutes in which to make the run to Rothbury before the evening passenger left Rothbury at 4-25 p.m. When traffic requirements made this timing unlikely, wagons to shunt at Longwitton and/or Ewesley, then the goods had to remain at Scot's Gap until after the arrival of the passenger from Rothbury at 4-57 p.m. In these circumstances the goods would not get back to Rothbury before 5-40 p.m. having left at 8-30 a.m. Good for overtime!

During the 1940s and especially in 1942 and '43, large numbers of soldiers were in training in the Rothbury area and all their stores and supplies were brought up to the Rothbury rail-head by train. This extra traffic increased the size of goods trains beyond what was normal for the line and in excess of the capacity of the ex-NER 0-6-0 (LNER class J21) tender engine stabled at Rothbury. On these occasions the goods had to wait at Scot's Gap until the arrival of the G5 tank from Rothbury after it had completed its evening passenger run. The sound of these two engines, the G5 piloting the J21, blasting up the 1 in 75 gradient to Longwitton with 25 or 30 fully laden vans on a calm summer's evening, as we heard it at the campsite, will long be remembered.

The reference to the sound of engine exhausts recalls the peculiarity, which I noticed with the G5's, as they changed "gear". On leaving a station there would be a series of 15 to 20 exhaust beats, then a pause while the driver adjusted the reversing lever to increase the cut-off, then the puffing would continue at a slower rate even though the train was going faster.

Additional trains were also run at the time of the big sheep sales at Rothbury, from mid-August onwards. These trains were always run in the evening, sometimes with one engine, sometimes with two, but the G5 tank was always the pilot. They worked these trains through to Morpeth and returned to Rothbury as light engines, sometimes as late as 10 or 11 p.m.

Although special passenger trains were run at one time in connection with Rothbury Races this did not happen during the war years. The only special train run for the public was on the third Saturday in September to Bellingham for the North Tyne Agricultural Society's Show. Extra carriages would be worked up to Rothbury on the previous Thursday and Friday to form a six or seven coach train on the Saturday morning. At Scot's Gap the front two vehicles would go to Morpeth as usual while the rear four or five would join the single carriage from Reedsmouth and proceed to Bellingham.

I have been reminded of the circular excursions that were run in the 1930s from Newcastle to Morpeth, then to Scot's Gap and Rothbury where the stop was prolonged so that passengers could obtain lunch. The train then returned to Scot's Gap and reversed to continue to Reedsmouth, Hexham and so back to Newcastle. One correspondent remembers that two of these trains were composed of four old dining cars with a six-wheeled luggage/brake van at each end and hauled by two J21 0-6-0's. He tells me that on another occasion the twin articulated 12-cylinder Sentinel steam railcar "Phenomena" was used but, whether this was a try-out to see how it performed on the line or because other stock was not available, is not known. This railcar was also used once a year for the railwaymen's annual outing from Blyth to Rothbury.

The prolonged depression of the 1930s caused great distress in the industrial areas of Tyneside and North Durham. Many people could not afford to buy the usual kinds of meat such as beef, lamb, mutton or pork and in those days chicken was a luxury. At the same time, much of Northumberland was over-run with rabbits. Put these two together and there was a relatively cheap source of meat that had only to be transported from country to town. The Rothbury line benefitted from the high rabbit population of the Wansbeck valley because a regular traffic developed at Middleton North, Angerton and Meldon stations of large wicker baskets filled with freshly killed rabbits as passenger train traffic to Newcastle and surrounding towns. The introduction of myxomatosis has since eliminated the demand for wild rabbits as meat.

The LNER made a final effort to improve the service in 1946 when a late train was run on Saturday evenings. This left Rothbury at 8-50 p.m. and called at all stations to arrive at Morpeth at 9-45 p.m. The return train waited until the pubs closed in Morpeth at 10-00 p.m. and left at 10-50 p.m. It should be remembered that at that time a large part of Northumberland was "dry". There wasn't a pub between Dyke Neuk, between Mitford and Meldon, and Rothbury. This late train called at all stations to Scot's Gap then express to Rothbury, arriving at 11-52 p.m. This must have been the latest ever time-tabled train to arrive at Rothbury.

After nationalisation the late train was withdrawn as was the afternoon Saturday train so that, by 1951, there were only two passenger trains a day in each direction, Mondays to Saturdays, and these were reduced to one brake/composite coach.

Passenger services were withdrawn in September 1952 but goods trains continued to run and a number of scenic excursions used the line in 1953, 1955 and 1956.

The Rothbury Line may never have hit the headlines of the railway press and, indeed, many people may never have known of its existence, but it was a line of great character. It was a joy to travel upon it and observe the devotion and commitment of the men and women who worked it. Until closure in 1952, it gave an atmosphere of reliance and dependability to an area that was sparsely populated and otherwise isolated from centres of social activity. The goods train continued to trundle to Rothbury until 1963 but then the line was closed and the track completely lifted by the Summer of 1964.

The last passenger train to run was the Railway Correspondence and Travel Society/Stephenson Locomotive Society special train named "Wansbeck Wanderer" on Saturday, 9th November 1963. Sadly, this train was not hauled by an ex-LNER locomotive, a G5 or J21, but by a foreigner – an Ivatt 4MT 2-6-0 No. 43129 – ex-LMS!

The Lead Lines

When, in 1923, the thirty or so independent railways in Great Britain were grouped into the Big Four – the London Midland and Scottish Railway, the London and North Eastern Railway, the Great Western Railway and the Southern Railway – the North Eastern Railway was the most successful and profitable constituent of the LNER. This was largely due to the amount of mineral traffic that it carried, especially in Northumberland and Durham.

Now it can be argued that, because coal is organic, it is not a mineral but this is the way in which it was classified in railway terminology. In 1923, 21½ million tons of coal were shipped from the Tyne alone and to this must be added the tonnages shipped from Sunderland, Seaham Harbour, Hartlepool and Middlesbrough. In addition, large quantities were taken out of the North-east coalfield by train. Coal contributed a very large proportion of the North Eastern Railway's mineral traffic with the remainder being shared between limestone, whinstone and iron ore.

The North Eastern Railway had nothing that could be remotely compared with the iron ore traffic of the Quebec, North Shore and Labrador Railway in Canada or the Hammersley Iron Ore Railway in Western Australia. Indeed, its iron ore traffic was limited to the relatively small quantities of ore mined in the Cleveland district of the North Riding of Yorkshire. So far as Northumberland was concerned, once coal was excluded the mineral wealth consisted of limestone and whinstone which provided some traffic on the Border Counties lines – but these were owned by the North British!

The only other mineral exploited in Northumberland and available to the North Eastern Railway was lead from the North Pennine ore-field based on Alston, Allendale, Weardale and upper Teesdale.

Lead had been extracted in the Alston area for many centuries but it wasn't until the latter part of the 18th Century that the industry grew and prospered. The London Quaker Lead Company leased the mineral rights on Alston Moor from Greenwich Hospital in 1750 with the official title of "The Governor and

Company for smelting down lead with pit coal and sea coal" This lease continued until 1882 when it was sold to the Nenthead and Tynedale Lead and Zinc Company. Four years later it was sold again to the Vielle Montagne Zinc Company – a Belgian concern. In neighbouring Allendale, mineral rights were taken up by the Blackett-Beaumont Company and in the next 100 years there was considerable competition between London Lead and Blackett-Beaumont in mining methods and extraction procedures at their smelt mills.

Mining activities expanded rapidly during the 19th Century in response to the increase in demand for lead for water pipes and house building for the large numbers of workers required to satisfy the demands of urban areas as a result of the industrial revolution. The London Lead Company centered their Alston Moor activities on the smelt mill at Nenthead while the Blackett-Beaumont Company had smelters at Allenheads and Allendale Town.

Transport in these remote areas was always a problem and it increased as the industry developed. At first there were no roads and lead had to be carried by pack horses in strings of 12-15, or more, to Stella Quay on the Tyne near Blaydon and to a smelt mill and refinery at Ryton. It was a slow method and very dependent on the weather. Both lead companies built a number of roads, lead roads, which can still be traced today on ordnance survey maps. These roads enabled horses and carts to replace pack horses for some, if not all, of the journeys. Even so, a horse-drawn cart could carry little more than half a ton. The magnitude of the problem may be better appreciated when it is realised that the first horse-drawn coach did not reach Alston until 1829.

The obvious solution, of course, was a railway but promoters were slow in getting off the mark. This was especially true in Allendale where the railway was not opened until lead mining was in decline. The promoters really "missed the boat". British lead mining was at its peak in the mid-1850s when the price of lead was £32 per ton. Thereafter, it gradually succumbed to cheaper lead imported from Spain and South America. The rise and fall of lead mining in the Northern Pennines is a fascinating story and readers are recommended to refer to books listed in the "Further Reading" section at the end of this book.

Alston and South Tynedale

The need for a railway to Alston was recognised as early as 1840 when the Newcastle and Carlisle Railway proposed a branch from their main line at Haltwhistle up the South Tyne Valley to Alston and then up the River Nent to Nenthead – the site of the London Lead Company's smelt mill. Although these proposals were enshrined in an Act of Parliament, work did not proceed.

In 1845, there was a proposal by the Wear Valley Railway Company to build a line from their Frosterley terminus to Alston, with a tunnel through Killhope, and on to Milton (Brampton Junction) and Carlisle. The 1846 depression put paid to that idea.

Photo: Ian S. Carr

Class 4MT 2-6-0 No. 43129 stands at Rothbury on 9th November 1963 with the RCTS/SLS "Wansbeck Wanderer", the last ever train to Rothbury. The line closed two days later.

Photo: Lens of Sutton

Haltwhistle station in its heyday. The Alston platform is the right side of the island and the viaduct over the South Tyne can just be seen at the extreme right of the picture.

Photo: Lens of Sutton

Lambley station and the viaduct in all its glory. The Lambley branch to join up with the Earl of Carlisle's railway can be seen going straight on from the station.

Photo: Lens of Sutton

Lambley station and staff. Note the check rails, the North Eastern slotted signal posts and the warehouse for small parcels. The smoke from the signal box chimney suggests that it was a cold day.

Photo: Lens of Sutton

Slaggyford station with its imposing signal box - but not a patch on the one now at Alston!

Photo: Robin Walton

Alston station in the early 1900s with a BTP tank bringing in a train of six-wheeled carriages.

Photo: Robin Walton

Alston goods yard in a hive of activity. This picture illustrates jobs for at least 25 people.

Photo: Lens of Sutton

Class J21 0-6-0 shunts at Alston while the snowplough placidly waits for something to do.

A new Bill was submitted in 1849. This terminated the line at Alston but added a westward spur at Lambley to an end-on junction with Lord Carlisle's railway from Brampton. This was the railway on which George Stephenson's "Rocket" finished its working life. The Royal Assent was given in July 1849 and construction commenced at both ends of the line in 1850. Goods trains ran from Haltwhistle to Shaft Hill (Coanwood) in March 1851 and from Alston and Lambley to Brampton Junction via Lord Carlisle's railway in January 1852. The delay in completing the line from Coanwood to Lambley was caused by the work required to build Lambley viaduct over the River South Tyne. This is a most impressive structure for a lowly branch line. There are nine 58 ft-span arches and seven 20 ft spans at a height of about 100 ft above the river. Alas, although it is a listed building, it is now in such a state of disrepair that it is unsafe to even walk over.

In 1864, there was another proposal involving Alston. This was to connect Alston with the South Durham and Lancashire Union Railway at Barnard Castle by a line up the South Tyne Valley to Tynehead, over Yad Moss and down Harwood and Upper Teesdale.

This was followed, in 1870, by a further proposal from the Wear Valley Railway. This time to link Alston to Stanhope. Eventually this line got as far as Wearhead, but not until 1895. Finally, in 1891, a plan emerged to link Alston to Middleton-in-Teesdale where the railway had arrived in 1868.

All these plans came to nought because of lack of support but, for a time, Alston was the railway promoter's "Mecca"

The line began at Haltwhistle at 380 ft and 13 miles later it arrived at Alston at 920 ft. It diverged in a south-easterly direction from the Newcastle-Carlisle line at the east end of Haltwhistle station and immediately crossed the River South Tyne by a 6-arched stone viaduct and proceeded on a rising gradient of 1 in 80 for ¾ mile to Plenmeller. The advent of the railway encouraged the sinking of a colliery at Plenmeller and by 1925 it was employing over 370 men and contributed a considerable amount of coal traffic to the line. It closed in 1932. There was a platform at the colliery and workmen's trains were run from both Haltwhistle and Alston.

Leaving Plenmeller the gradient increased to 1 in 70 for 1½ miles then eased to 1 in 100 for the next ¾ mile to Featherstone Park station. Like all the stations on the line it had a single platform with a level crossing at the south end. The sidings incorporated a run-round loop but this was not part of the main line and Featherstone Park was not a passing place nor a block post. There was a signal box but it was only a gate box for the crossing. The station had its moments of glory during the 1914-18 war when a military camp was established nearby and again from 1939-45 when, again, an army camp was set up in the grounds of Featherstone Castle while the castle itself was used to house prisoners of war.

Featherstone Colliery was situated about half a mile east of the railway and about ¾ mile south of Featherstone Park station. It must have been a relatively large one because it employed 228 people in 1899. However, by 1905 the main seam was worked out. The reception sidings between Featherstone Park and Coanwood stations were connected to the colliery by an inclined tramway. Coal was transferred from tramway tubs to railway wagons at a loading hopper in the siding complex.

Coanwood station, originally called Shaft Hill, was 4 miles from Haltwhistle on a gently falling gradient of 1 in 525. It served adjacent communities engaged in coal mining at Coanwood colliery and quarrying at the whinstone quarry half a mile further on. It was the first block post after leaving Haltwhistle and had a lengthy passing loop. However, since there was only one platform (on the down side), it was not possible for two passenger trains to pass. If an up passenger needed to pass a down goods then they would have to take the right-hand tracks. As at Featherstone Park, there was a level crossing and a station staff of two porter-signalmen although at one time Coanwood had its own station master.

Coanwood colliery sidings were ¼ mile south of the station and entered by facing points from Haltwhistle. The colliery was a very old one, dating back to 1780. The coming of the railway stimulated further development and a new 442 ft shaft was sunk in the early 1860s. The colliery had an eventful existence. At one time it was abandoned when it became filled with water. Mining was resumed after the workings had been pumped out and continued until 1917. The colliery and associated coke ovens employed about 50 people in 1910. The siding from the main line branched out into no fewer than 11 sidings to various parts of the colliery complex. I find it difficult to accept that coal could be mined, brought to the surface, washed, screened, coke ovens charged and emptied, and 11 sidings kept in use by a total work force of only fifty!

The whinstone quarry sidings were only a short distance beyond the colliery but were, nevertheless, protected by up and down signals. These were so close to Coanwood up home and Lambley down home that both posts carried distant signals as well. At its peak, in 1898, the quarry employed 55 men but it soon worked itself out because it was hemmed in by the river on one side and the railway on the other.

The approach to Lambley viaduct was through a cutting which curved to the right. This viaduct is the most spectacular engineering structure on the line and it is a great pity that no-one is prepared to provide money necessary for its upkeep. Eventually it will just fall down.

On leaving the viaduct, the railway takes a sharp curve to the left, over a trailing junction with the Lambley branch, and into Lambley station – $4^7/_8$ miles from Haltwhistle. It was one of the more interesting stations on the line because of the junction with the Lambley branch and its end-on junction with Lord Carlisle's railway. Although there was a small goods shed opposite the curved platform at Lambley station it was used only for parcels and small consignments. Wagon and van loads were dealt with at Lambley village, ¼ mile along the branch, where there was a goods siding.

Lambley was also a block post with the unusual arrangement that trains could pass if one of them was shunted into the branch!

South from Lambley, the railway climbed at gradients varying from 1 in 100 to 1 in 254 with a short descent at 1 in 264 at the eighth mile-post to cross the Knor Burn. The stone built viaduct has four 30 ft spans at a height of 47 ft above the burn. There are two other attractive viaducts between Lambley and Knor Burn. The first crosses the Glendue Burn, a mile south of Lambley, in five 20 ft span arches. A mile and a half further on is the Thinhope Burn with five 30 ft arches crossing the burn and one 39 ft arch crossing the road. The viaduct is locally known as Burnstones because of its proximity to the hamlet of that name.

After crossing Know Burn the railway returned to a rising gradient of 1 in 155 and in half a mile reached Slaggyford station. The track layout included a loop but this was not used for trains to pass. A loading dock and a small goods shed were situated on this loop which extended into a long dead-end siding running south parallel to the main line. Another siding with facing points from the north ran behind the goods shed and terminated in a typical North Eastern Railway coal drop. Like Featherstone Park and Coanwood, Slaggyford had a level crossing protected by signals operated from a signal box which had a wheel to work the crossing gates. In 1916, a private siding of some considerable length was laid south of the station to serve the Anthracite Coal and Lime Company. Coal was loaded into railway wagons from an overhead gantry at the end of the siding. Access to the siding was controlled by a ground frame.

About a mile south of Slaggyford the line is crossed by the road from Brampton to Alston and then crosses Thornhope Burn on a viaduct of four 30 ft. arches 45 ft above the ground. This viaduct is also known as Lintley after the name of the farm adjacent to it. South of the viaduct, a ground frame gave access to Barhaugh Colliery sidings, again with facing points to the north. These sidings had a run-round loop with catch points to prevent breakaways getting on to the main line. Barhaugh colliery was situated about half a mile away on the opposite side of the river and coal was conveyed by an aerial ropeway from the colliery to the siding. It was a small colliery, opened in 1909 and employing about 14 men. It was worked until 1939.

The steepest part of the line with a gradient of 1 in 56 began after Barhaugh siding and continued for a mile through the hamlet of Kirkhaugh before falling at 1 in 528 to cross the Lort Burn. The viaduct here has three arches, one of 20 ft span and 34 ft above the burn and a small one at either side. To the east of Lort Burn and standing on the haugh, or river-side plain, is Kirkhaugh Church with its beautiful slender spire. Although apparently in the middle of nowhere it remains in use for regular services and hosts a number of special musical events each summer.

After Lort Burn there is a mile of straight track on a rising gradient of 1 in 176 to Gilderdale Burn – the boundary between Northumberland and Cumbria. Here is another three-arched viaduct with a centre span of 49 ft and a 29 ft span at each side, also 34 ft above the burn.

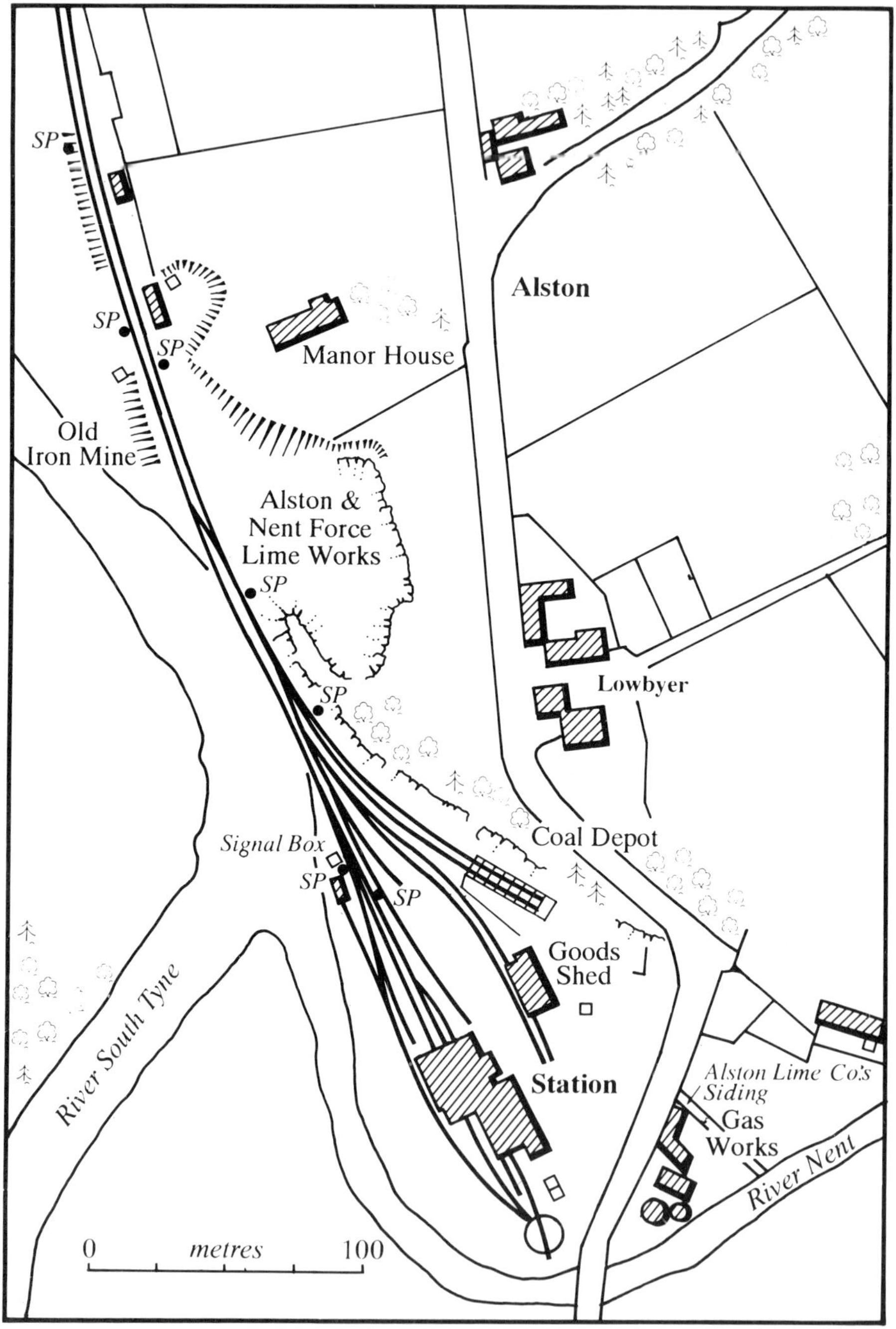

Alston station

The final approach to Alston involves 1¼ miles of gentle gradients of 1 in 264 easing to 1 in 396. The River South Tyne is crossed by a substantial stone viaduct of three 39 ft arches, half a mile from the terminus.

Alston station was a gem of a branch line terminus. It had everything. Substantial and attractive stone station buildings with house attached; an overall roof that covered the single platform and an adjacent carriage siding; a run-through shed for two engines; a workshop equipped with lathes and a smithy and a 7,000 gallon water tank built into the roof; a coaling-stage; and, at the south end of the platform, a 42 ft turn-table which gave access to the engine shed and run-round lines. There was even a shed to house a snow plough but this building had gone by 1930. The signal box was at the north end of the station between the run-round line and the river. It was a typical North Eastern box with brick base and a hip-roof.

Facilities for goods traffic were extensive with special provision being made for the loading of stone and other minerals. There was a large goods shed, 70 ft × 40 ft, with an internal loading platform that enabled road vehicles to be loaded under cover. Two sidings served a double track coal depot with eight cells under each track. Coal and coke were important imports at Alston for not only were they the main source of heat for the 4,000 or so population but coal was used at the gas works and coke at Alston Foundry, both on the opposite side of Hexham road to the station. The triangular area between the goods siding behind the station platform and another one, parallel to the siding through the goods shed, was built up to a height of 4½ ft above rail level so that horse-drawn carts could tip directly into railway wagons. A cattle dock was situated between these two sidings. Another siding branched off the goods shed siding and went through a tunnel under Hexham road to a private stone yard. In all, sidings at Alston station could accommodate 152 four-wheeled railway wagons.

North of the station and on the east side of the line there were four more sidings that could hold another 150 wagons. They were used by the Alston and Nentforce Quarry Company and the Vielle Montagne Zinc Company with limestone, ores and other minerals being loaded from an aerial flight from Blaghill quarry, a 2 ft gauge self-acting incline from Newshield quarry (both owned by Alston Lime and Coal Company), and from horse-drawn carts over loading banks.

It will be apparent from the above that Alston station was a very busy place in the early part of the 20th Century. An old photograph taken about 1912 shows activities in progress that gave employment to upwards of 50 men and women working at the station and on the railway, at the coal depot and as carters and labourers.

The line was fully signalled throughout with single line working controlled by electric tablet from Haltwhistle to Coanwood, train staff from Coanwood to Lambley and electric tablet from Lambley to Alston.

Signalling arrangements between Haltwhistle and Alston when the line was at its zenith are summarised as follows:-

Miles from Haltwhistle	*Location*	*Apparatus*
1.0	Plenmeller Colliery siding	Ground frame, tablet release
3.1	Featherstone Park	Gate box, signals. Tablet controlled frame for sidings
3.6	Featherstone Colliery sidings	Ground frame, tablet release
4.2	Coanwood	Signal box, block post
4.4	Coanwood whinstone sidings	Dwarf frame (6 levers) with signals released by staff key
4.8	Lambley	Signal box, block post
8.6	Slaggyford	Gate box, signals, tablet release
9.8	Barhaugh Colliery sidings	Ground frame, tablet release
12.9	Alston	Signal box, block post

When, in 1852, the line was opened throughout from Haltwhistle to Alston there were three passenger trains a day each way. The first train left Alston at 7-25 a.m. and the last one got back at 7-48 p.m. The journey time was 40 minutes for the 13 miles.

In 1862, the Newcastle and Carlisle Railway, owners of the Alston line, was absorbed by the North Eastern Railway who proceeded to develop it into a typical North Eastern branch line with well developed facilities for coal and mineral traffic. By 1884, the three passenger trains had been augmented by a fourth one (in the afternoon) on Saturdays and the journey time was reduced to 35 minutes.

The period of North Eastern ownership saw the decline and virtual disappearance of lead mining. In 1858, Alston Moor produced 5,382 tons of lead ore which yielded 3, 960 tons of lead and 28,056 ozs of silver. The Nenthead smelt mill closed in 1882 and this was the end of lead in any quantity. However this was more than compensated by the growth in the extraction of zinc ore, from 1,916 tons in 1883 to 9,873 tons in 1914, and by the increase in coal mining, limestone quarrying and the extraction of fluospar and other minerals. In 1895, for example, 48,566 tons of coal, coke, lime and limestone were received at Lambley, most of it from collieries, coke-ovens and lime works on Tindale Fell. There was also a smelter at Hartleyburn. This volume of traffic is surprising because Lord Carlisle's railway, over which it came, had its own access to Brampton Junction by means of the incline (or decline!) from Hallbankgate. In addition to this traffic there was, of course, coal and coke from Coanwood Colliery and coal from Featherstone Colliery. Again, in 1914, Alston loaded 9,873 tons of zinc ore as well as large quantities of limestone and other minerals.

These changes in mining activity promoted an additional passenger train for workmen. In 1921 a workmen's train left Alston at the un-earthly time of 3-10 a.m. and arrived at Haltwhistle at 3-50 a.m. after calling at the halt at Plenmeller Colliery. For some reason it ran an hour earlier on Saturdays. The return working departed from Haltwhistle as an empty train at 3-58 a.m., picked up miners at Plenmeller and called at all stations to arrive at Alston at 4-34 a.m.

1921 was the last year that this train was advertised in the public time-table for, by July 1922, it had disappeared even though Plenmeller was at its peak of production employing 292 underground workers and 80 on the surface. Perhaps the train continued to run as a "ghost" train. The publicised service was for four trains a day with the last one leaving Haltwhistle at 8-00 p.m. – 15 minutes later on Saturdays.

Goods workings in 1921 show that they were done by Carlisle crews because a light engine and van left Carlisle at 4-45 a.m. and arrived at Alston at 6-10 a.m. They left Alston at 6-45 a.m. with a cattle train which ran through to Carlisle to arrive at 9-29 a.m. Another goods left Carlisle at 5-35 a.m. and, after shunting at Haltwhistle, departed for Alston at 7-47 a.m. It ran non-stop and arrived at 8-22 a.m. After shunting it also departed as a cattle train and, apart from stopping at Featherstone Park to pin down the brakes, it ran direct to Haltwhistle to arrive at 10-15 a.m. The time-table acknowledges that this train ran in connection with Haltwhistle stock sales. One strongly suspects that both these trains ran only once a week, probably Monday, and not necessarily throughout the year.

The regular pick-up goods left Carlisle at 8-11 a.m. and arrived at Haltwhistle at 9-32 a.m. After an hour there it left at 10-36 a.m., passed by Plenmeller Colliery and arrived at Featherstone Park for a four minute stop at 10-48 a.m. Coanwood arrival time was 11-01 a.m. and it spent 23 minutes there before proceeding to Lambley. Fifty-four minutes were allowed to shunt the Lambley branch and this would include a trip to the end of the line at Lambley Fell Colliery. Leaving Lambley at 12-22 p.m., this train was due at Slaggyford at 12-34 p.m. with 18 minutes allowed to shunt the sidings before setting off to arrive at Alston at 1-04 p.m. – 4 hours and 19 minutes after leaving Carlisle. I hope that, on reading this, opponents of the nationalised British Rail will realise how fast and efficient the privately owned railway companies were!

The return working departed Alston at 2-05 p.m. and called at Slaggyford at 2-20 p.m. when 10 minutes were allowed to shunt Barhaugh Colliery sidings. Twelve minutes, from 2-43 p.m. to 2-55 p.m., were allowed at Lambley, 10 minutes at Coanwood and 10 minutes at Featherstone Park, presumably to pick up loaded wagons martialled on the outward journey. Haltwhistle was reached at 3-40 p.m. and then this Alston pick-up goods train set off at 4-25 p.m. for Blaydon sidings where it arrived at 8-06 p.m. The identity and location of the engine and crew involved in this eastward jaunt is not revealed but it took them 3 hours and 45 minutes to cover the 33¼ miles from Haltwhistle to Blaydon! The working time-table makes no reference to coal traffic from Plenmeller so how and when this was worked remains a mystery.

The formation of the LNER in 1923 had little effect on the Alston branch. Engines were the same, carriages were the same but eventually changed colour from dark maroon to drab brown, and signals and track were the same. Deterioration in mining and quarrying continued so that, by 1938, Plenmeller Colliery had closed, Barhaugh Colliery had closed and traffic at Lambley was limited to coal from Lambley Fell Colliery. One goods train a day was able to meet all requirements and this left Haltwhistle at 9-45 a.m., called only at

HALTWHISTLE and ALSTON

Weekdays (Down)

Distance from Haltwhistle	DOWN	1	2	3	6	8	9	12	14		15	16	17	18	19	20	21	22	23	24	25	26	27	28
								PASSENGER	D Goods		PASSENGER		PASSENGER			PASSENGER	PASSENGER	PASSENGER	PASSENGER	PASSENGER		PASSENGER	PASSENGER	
																S O	SX	S O	SX	S O		SX	S O	
									arr.	dep.														
M.C.								a.m.	a.m	a.m.	a.m.		p.m.			p.m.	p.m	p.m.	p.m.	p.m.		p.m	p.m.	
...	**Haltwhistle** ⊕							8 10	—	9 45	11 35		1 35			3 35	5 5	6 13	6 20	7 45		9 10	10 42	
3 0	Featherstone Park							8 18	..	....	11 43		1 43			3 43	5 13	6 21	6 28	7 53		9 18	10 50	
4 8	Coanwood ⊕ +							8 22	..	.	11 47		1 47			3 47	5 17	6 25	6 32	7 57		9 22	10 54	
4 56	Lambley + ⊕							8 25	....	*	11 50		1 50			3 50	5 20	6 28	6 35	8 0		9 25	10 57	
8 41	Slaggyford							8 33	..	..	11 58		1 58			3 58	5 28	6 36	6 43	8 8		9 33	11 5	
13 0	**Alston** ⊕							8 42	11 17	—	12 7		2 7			4 7	5 37	6 45	6 52	8 17		9 42	11 14	

Sundays (Down)

1	2	3	4	5

Weekdays (Up)

Distance from Alston	UP	1	2	7	8	9	10	11	12	13	14	16	17	18	19	20	21	22	23	24	25	26
						PASSENGER		PASSENGER		PASSENGER		D Goods	PASSENGER	PASSENGER	D Goods	PASSENGER	PASSENGER		PASSENGER		PASSENGER	
												SX	S O	SX	S O	S O			SX		S O	
																			J			
M.C.						a.m.		a.m.		p.m.		p.m.	p.m.	p.m.	p.m.	p.m.	p.m.		p.m.		p.m.	
....	**Alston** ⊕					7 10		10 15		12 35		2 30	2 45	3 55	4 40	5 30	7 0		8 20		9 10	
4 39	Slaggyford					7 19		10 24		12 44		*	2 54	4 4	*	5 39	7 9		8 29		9 19	
8 24	Lambley ⊕ +					7 27		10 32		12 52		*	3 2	4 12	*	5 47	7 17		8 37		9 27	
8 72	Coanwood + ⊕					7 30		10 35		12 55		*	3 5	4 15	*	5 50	7 20		8 40		9 30	
10 0	Featherstone Park					7 34		10 39		12 59		b	3 9	4 19	b	5 54	7 24		8 44		9 34	
13 0	**Haltwhistle** ⊕					7 41		10 46		1 6		3 45	3 16	4 26	5 55	6 1	7 31		8 51		9 41	

Sundays (Up)

1	2	3	4	5

J—Carlisle arr. 9.25 p.m. p. 145.

b—Brakes.

D 20 m.p.h.

4

From the L.N.E.R. (North Eastern Area) "Working Timetable", September 1938

Photo: Lens of Sutton

Class G5 0-4-4T No. 67315 stands at Alston in the 1950s. The station is still intact with overall roof and engine shed - but not for long.

Photo: Ken Taylor

Hexham station when three platforms were in use plus the bay on the right for trains to Allendale. Note the large number of cattle wagons to serve the needs of Hexham and Tynedale auction marts.

Photo: Lens of Sutton

Langley station with two platforms but no sign of any signals. All passenger trains used the platform on the right, the other was on a passing loop for goods and parcels controlled by two ground frames.

Photo: J. Foster-Smith

Staward station in January 1951 with the demolition men in. The loop line has already been lifted.

Photo: Lens of Sutton

Allendale station as built in 1868 with a short platform and starting signal.

Photo: Lens of Sutton

Allendale station after improvements in 1884. There is now a longer platform with lights, and a glass covered waiting shelter, but the signal has gone.

Photo: Northumberland County Record Office

Alnwick station in 1935. The grand scale of this station and its elegant style were obviously influenced by the fact that Alnwick is the seat of the Duke of Northumberland.

Photo: Northumberland County Record Office

Edlingham viaduct looking towards Edlingham station with the Alnwick to Rothbury road on the hill-side in the distance.

Lambley and arrived at Alston at 11.17 a.m. After shunting and simmering at Alston, it left at 2-30 p.m. (4-40 on Saturdays) and called at all stations as required to arrive at Haltwhistle at 3-45 p.m. (5-38 Saturdays).

Passenger workings were interesting in that under the management of the LNER , they increased to six a day with seven on Saturdays. The last train from Haltwhistle ran as late as 10-42 p.m. for the Saturday night revellers. From Mondays to Fridays the 8-20 p.m. train from Alston was a through train, express from Haltwhistle to Carlisle, arriving there at 9-25 p.m. This must have been one of the few through tains ever to operate. The balancing return working from Haltwhistle at 9-10 p.m. used the carriages of a stopping train which left Carlisle at 5-40 p.m. and terminated at Haltwhistle at 6-56 p.m. On Saturdays this train returned to Carlisle at 9-32 p.m.

Much has been written elsewhere (see "Further Reading"section) about the engines used on the Alston line. My own personal recollections are of G5 0-4-4 and A8 4-6-2- tank engines on passenger trains and J21 and J39 tender engines on goods trains. My introduction to the line was rather late in my Northumberland railway experience and didn't occur until 1946. We stopped for a picnic on the narrow road running up the east side of the valley and overlooking the railway between Lambley and Slaggyford. It was a Bank Holiday Monday, whether Whit or August I cannot remember, but trains seemed to be passing by about every half hour. The line was busy. A G5, too far away to make out its number, shared the service with an A8 and while the former headed four compartment carriages the latter hauled five of mixed clerestory and eliptical roofed stock – all of North Eastern origin. I have since endeavoured to find out why the line was so busy that day but without success.

Subsequently I saw more of the line and the excellent work done on it by the ubiquitous G5 tanks. They were great engines and it is a great loss to railway history that none has been preserved. I remember excursion trains parked in the stone sidings at Alston with a J39 or a K1 on the front and my last poignant memory of steam on the line was in 1960. We were on our way by road from Barnard Castle to Newton Stewart when, on passing Lambley Fell Colliery, we saw a BR 3MT 2-6-0 standing with 8 loaded 21 ton steel hopper wagons waiting to return to Lambley Junction. This must have been one of the last trains on the Lambley branch.

Closure of the line was threatened in 1959 but British Rail, under pressure from the Transport Users' Consultative Committee, agreed to keep it open because of Alston's isolated position and the ease with which roads were blocked by snowstorms. Steam hauled passenger trains were replaced by two-car Metropolitan-Cammel D.M.U's but these were based at South Gosforth where they had to go to re-fuel. Consequently, savings created by getting rid of steam were squandered by empty stock workings between Haltwhistle and Hexham and South Gosforth at the beginning and end of every day.

Goods traffic ceased in 1965 and the line became a one-engine-in-steam long siding from Haltwhistle with a run round loop retained at Alston controlled by a key-operated ground frame.

After further successful efforts by the T.U.C.C. to keep the line open it was finally decided that, instead of spending £73,000 a year to make good the so-called running loss, the Government would spend £300,000 on the construction of a new "all-weather" road between Coanwood and Lambley with a new bridge across the river. This road took three years to build and, when it was completed in the Spring of 1976, the railway was closed. I did not travel on the last train on 1st May 1976, an 8-car D.M.U. from Carlisle to Alston and back organised by Border Round Table, but I did make the return journey from Alston to Haltwhistle during the previous week. It was a depressing experience. Gone was the overall roof at Alston station, gone were the engine shed, turntable, signals and the sidings in the goods yard. There were eight of us on the 13-00 hours from Alston and six more were picked up on the journey. All the stations were unstaffed and derelict, level crossings were open to road traffic at Slaggyford, Coanwood and Featherstone Park with the train having to stop before crossing the road. The return journey from Haltwhistle at 15-45 hours began with twenty-two passengers, twelve of whom completed the journey to Alston which, incidentally, took 34 minutes – one minute less than in 1884!

Meanwhile, the "all-weather" road is a joke. Fortunately we have not had any Winters recently to compare with those of 1947 and 1963 but, when we do get three or four inches of snow, Alston is cut-off from the outside world.

The Alston line was built 124 years earlier to carry lead from the North Pennine ore-field. It missed the boat because by the time it was opened lead mining was in decline. However the railway did carry a lot of coal, zinc ore, limestone, lime and fluospar and this probably justified the costs of construction. Not so the other lead line up Allendale.

Allendale

Lead mining in the Allendales, both East and West, was controlled by the Blackett-Beaumont Company which held leases from Greenwich Hospital as early as 1692. They had smelters at Allendale Town and Allenheads and for 30 years the output of these smelters was carried by packhorse or horse and cart to Haydon Bridge station for onward carriage to Ryton by train. The cost of this transport was relatively high and became critical in 1860 when lead began to be imported from Spain with a consequent fall in the price of the home produced metal. It became essential to reduce costs as much as possible and T. J. Bewicke, the manager of the Beaumont Company, determined that a railway was the answer. There was another smelt mill in the area, at Langley, operated by the London Lead Company and they gave their support to Bewicke's proposal. The North Eastern Railway was also sympathetic to the idea and even offered to subscribe £10,000 towards the cost provided that there was strong local support.

A route was surveyed from Hexham through Langley to Allendale and up the East Allen valley to Allenheads. There was no opposition to the Bill when it was presented to Parliament and the Hexham and Allendale Railway Act received the Royal Assent in 1865. It was in two parts, a 13 mile line from Border Counties Junction to Allendale Town, and a 7 mile extension to Allenheads. The second part was never built and the first part only got as far as Catton Road, ¾ mile short of Allendale Town but only ¼ mile from the Allen smelting mill. The section from Border Counties Junction to Langley was opened in August 1867 and the remainder to Catton Road in January 1868. The name of the terminus was changed from Catton Road to Allendale in 1898 when it was obvious that any extension to Allendale Town would never be built.

The line began at Border Counties Junction, one mile west of Hexham, and the turn-out was immediately west of the junction with the North British's Border Counties line. On negotiating the points, engines were faced with a fearsome gradient of 1 in 40 from 150 ft to 800 ft at Langley, a distance of about 7½ miles. Three miles from Border Counties Junction there was a siding (facing east) into Glen Dhu quarry where stone was brought to railway wagons by an internal tramway. The line curved south after Glen Dhu and two miles on was the single platform station at Elrington. It had a single turn-out, facing east, controlled by a ground frame. It branched into two sidings and had a long head-shunt clear of the main line. Continuing the climb, the line curved south-west for 2½ miles to Langley station. This station had two platforms although one was used for only goods and parcels traffic. The double track provided a passing loop that was controlled by a ground frame at each end.

The smelt mill at Langley was built by Greenwich Hospital, in 1767, and operated by the London Lead Company. Over the years it grew "like Topsy" in an unplanned and haphazard way until it became the size of a village. One of its distinguished features was an overground flue which carried smoke and vapours from the smelting process to a chimney half a mile to the east and at a height of 975 ft. When the railway was built, this flue had to be carried over the line by a bridge and smelting was suspended while this bridge was built. The flue can still be seen from the road between Cartsbog and Stublick and the chimney has recently been repaired and "made good".

Railway access to the smelter was a siding about 600 yds east of Langley station. As with all the other sidings on the line, it was controlled by a ground frame locked by the train staff.

Langley, at 800 ft, was the summit of the line and from there it fell quite steeply at 1 in 50 to the Cartsbog road and then almost level to Staward station, 1¾ miles from Langley. When built, Staward had two platforms and a passing loop. The station house was stone built and there was a small goods yard with coal cells. It was the nearest station to Whitfield and served the West Allen valley.

The final 2¾ miles to Allendale station was a gentle descent in a south-easterly direction high above the River East Allen. The station had one passenger platform and a 42 ft turn-table at the end of it gave access to the run-round loop. A siding continued from the east end of the loop to a set of coal cells adjacent to Catton Road. At the west end of the station a turnout branched into three with the centre one serving a small goods shed. Allendale Creamery was immediately north of the goods yard but it didn't have its own siding.

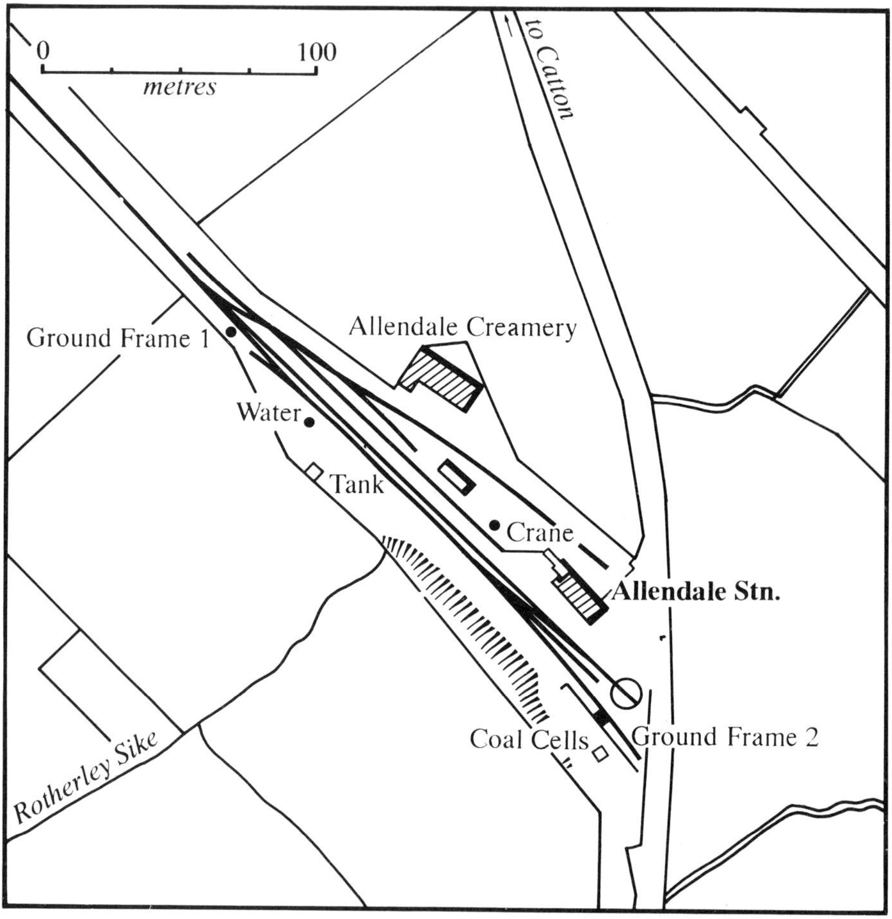

Allandale station

There was no signal box at Allendale. Points were operated from two ground frames, one at each end of the station. These were locked by a key on the train staff. Early photographs show a signal at the west end of the platform but this disappeared when the station was enlarged in 1884.

When the line opened in 1868, the population of Allendale parish was about 6,000 and there were three passenger trains a day, Mondays to Saturdays, with an extra one on Tuesday (Hexham market day). The service was worked from Hexham with the first train leaving at 7-55 a.m. and arriving at Allendale at 8-30 a.m. By 1911 the population had fallen to 2,000 and only one important mine was still working.

In spite of the population decline the service remained the same and, in 1922, there were trains from Hexham at 7-35 a.m., 2-15 p.m. and 6-05 p.m. with an additional one on Tuesdays at 3-50 p.m. Allendale departures were at 8-25 a.m., 3-05 p.m. and 7-02 p.m. plus one at 4-45 p.m. on Tuesdays. The large gap between the 8-55 a.m. arrival at Hexham and the 2-15 p.m. departure for Allendale was presumably to allow the daily goods train to go to Allendale and back, though with passing loops at Langley and Staward this should not have been necessary.

The growth in popularity of buses and the flexible service that they provided rendered the railway impotent to the challenge and passenger services were withdrawn on 22nd September 1930. Consequently, I never had an opportunity of seeing the line in full operation or of travelling on it. I understand that trains were worked by tank engines, mostly B.T.P. types, in the early part of the century and then by G5's for which the permitted load was 105 tons.

The goods and parcels service continued to operate and, in 1938, there was a No. 1 express goods from Hexham at about 7-30 a.m. which conveyed empty milk cans, parcels, newspapers and mails. This returned from Allendale at about 8-30 a.m. with milk, parcels and livestock. A class D goods left Hexham at about 10-30 a.m. and this also conveyed any parcels traffic. It returned as, and when, traffic permitted. There was also a more specifically termed No. 1 express parcels which left Hexham at 6-50 p.m. and arrived at Allendale at 7-25 p.m. The return times were 7-40 p.m. from Allendale with arrival at Hexham at 8-15 p.m. This train ran only from Mondays to Fridays. On Saturdays a class B goods left Hexham at 5-52 p.m. and got back at 7-40 p.m. I find it extremely difficult to believe that the Allendale branch could generate enough traffic to justify three goods and parcels trains a day but this was the way in which private enterprise, in the form of the LNER, operated.

I was evacuated to Hexham in 1940 and my personal recollections of the line was of J21's storming up the bank from Border Counties Junction with a train of vans containing animal feeding stuffs and coal wagons. Much of the ascent was in cuttings which amplified the bark of the engine exhaust very effectively. We used to explore the countryside around Hexham on push-bikes and this frequently included a ride to Low Gate and then down to Elrington to see the morning goods slowly descending the 1 in 40 back to Hexham.

HEXHAM and ALLENDALE

Distance from Hexham	DOWN	Weekdays 1	2	3	4	5	6	7	8	9	10	12	13	14	15	Sundays 1	2	3	4
					NO. 1 EXPRESS GOODS	D Goods			B Goods					NO. 1 EXPRESS PARCELS					
									SO					SX					
					PY	VJ													
M. C.					a.m	a.m.			p.m.					p.m.					
.....	**Hexham**				*	*			5 52					6 50					
1 10	*Border Counties* .. ✝				.....	.....			.. .					6 55					
6 25	Elrington Halt........				*	*			..					..					
8 68	Langley-on-Tyne ..				*	*			..					.....					
10 55	Staward				*	*			..					..					
13 35	**Allendale** .. ✝				DD	DD			6 38					7 25					

Distance from Allendale	UP	Weekdays 1	2	3	4	5	6	7	8	9	10	11	12	Sundays 1	2	3	4
					NO 1 EXPRESS GOODS	D Goods				B Goods		No. 1 EXPRESS PARCELS					
										SO		SX					
					MY	VJ				A							
M. C.					a.m.	p.m.				p.m.		p.m.					
.....	**Allendale**.... ......✝				U	U				6 55		7 40					
2 60	Staward .. – –				*	*				*		..					
4 47	Langley-on-Tyne				*	*				*		.....					
7 10	Elrington Halt				*	*				*		..					
12 25	*Border Counties*.....✝				.....	.....				7 36		8 10					
13 35	**Hexham**				DD	DD				7 40		8 15					

A—Stops at Bishopside Halt between Allendale and Staward for milk traffic. J—Conveys parcels traffic if required. K—Conveys Milk Van and parcels, also empty wagons from Staward and Langley-on-Tyne. M - Conveys milk and parcels traffic, also live stock. P—Conveys empty milk cans, parcels, newspapers, and mails. V—Worked by engine and men of 7.20 a.m. Blaydon to Hexham. Y—Worked by engine and men of 5.30 a.m. Blaydon to Hexham.

4

B 25 m.p.h. D 20 m.p.h.

From the L.N.E.R. (North Eastern Area) "Working Timetable", September 1938

Signalling of the line was very simple and based on the one-engine-in-steam principle with ground frames operated by the train staff at all sidings. Details are as follows:-

Miles from Border Counties	*Location*	*Apparatus*
5.1	Elrington	Ground frame
7.6	Langley No. 1	Ground frames
7.8	Langley No. 2	
9.3	Staward No. 1	Ground frames
9.4	Staward No. 2	
12.2	Allendale	Two ground frames

After the war, road transport took what traffic remained, mainly milk and livestock coming out of the dale and coal, fertiliser and animal feed going in. Closure was inevitable and the last goods train ran on 20th November 1950.

The Hexham and Allendale Railway was built too late to benefit from the rich mining activity that had once taken place in the valleys of the East and West Allen. The railway opened in 1868 but, by 1870, lead mining was in decline and the 1,000 tons of lead carried that year fell to 100 tons by 1900. Other traffic from Allendale was milk (until the Milk Marketing Board directed it all to lorries) and livestock, while a limited amount of timber was loaded at Staward and Langley from extensive woods round about.

In retrospect, it is probable that the line never made a profit and should never have been built when it was. If it is true to say that the Alston Line "missed the boat", then the Allendale Line "arrived at the quay after the boat had disappeared over the horizon".

So ends the sad tale of "The Lead Lines".

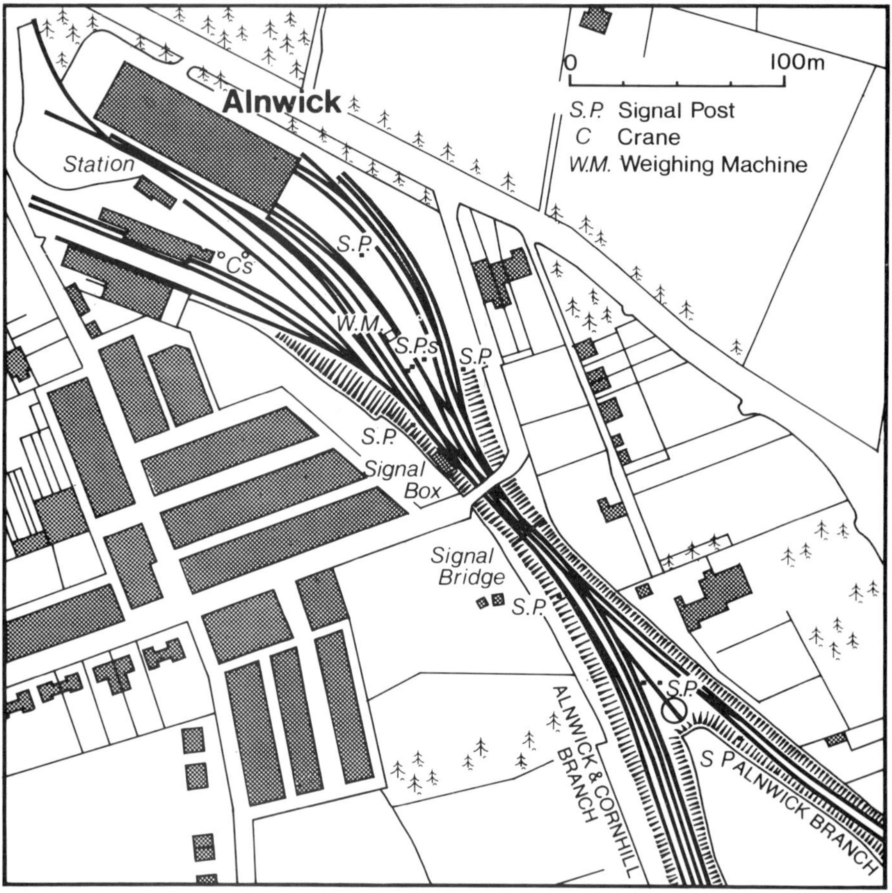

Alnwick station

Photo: Northumberland County Record Office
Whittingham station. The only island platform on the Wooler line.

Photo: Real Photographs Collection/ Reproduced Courtesy of Ian Allan Publishing
One of the engines that I used to watch! Class G5 0-4-4T No. 1762 at Wooler in 1936.

Photo: Stuart Sellar
Ivatt Class 2 2-6-0 No. 46482 at Wooler North, 29th March 1965. Note the delapidated state of the signal box.

Photo: W. S. Sellar
Ivatt Class 2MT 2-6-0 stands in Wooler station shortly before the withdrawal of goods services in 1965.

The Wooler Line

My earliest recollection of railways goes back to family holidays at Wooler in 1935 and 1936. The highlight of those days was the arrival of two daily goods trains, one from Alnwick and the other from Tweedmouth via Coldstream. Consequently, it became a burning childhood ambition of mine to travel over the Wooler line but, because the passenger service had been withdrawn in 1930, there was very little hope of this ambition being realised. There was, however, one faint chance. In those days the LNER had camping coaches in a siding at Akeld and elsewhere on the line and it was a condition of occupancy that the tenants of these coaches travelled by train. Hence, when camping coaches were located on lines from which passenger services had been withdrawn, the railway was obliged to provide travel facilities for their camping coach tenants. In the case of the Wooler line this was done by adding a third class carriage to the daily parcels train, usually on a Saturday. How I envied the passengers boarding this carriage at Alnwick for the journey to Akeld. I recall using every possible argument in an effort to persuade my parents that what the family really needed was a holiday in a camping coach at a station on the Wooler line. Alas, without success, because my Mother insisted that it would be no holiday for her if she still had to do the cooking. So we ended up in lodgings at Wooler. Consequently, now that the line has gone, one of many childhood ambitions is unfulfilled.

The Wooler line, or, to give its official LNER title, the Coldstream Branch, was built to fill the gap that was left when the Northumberland Central Railway failed to complete the line for which it received an Act of Parliament in July 1863. This railway, the Northumberland Central, was to run from Coldstream, on the North Eastern Railway's Berwick and Kelso branch, to Crookham, Akeld, Wooler, Wooperton, and Rothbury to connect at Scot's Gap with the Wansbeck Valley Railway.

As so often happened in the promotion of railways, talk was the cheapest and most active part of the proposal. When the call was made for hard cash to enable the proposal to be carried out the money was not forthcoming. The result was that the Northumberland Central Railway became no more than a short 13 mile branch line from Scot's Gap to Rothbury that was completed in November 1870.

This left landowners in north Northumberland without the benefit of a railway to link them to other parts of the country. In 1881, new proposals were submitted for a Central Northumberland Railway which would be in two parts. One section would be from Newcastle to Scot's Gap via Ponteland and Belsay and the other from Rothbury to Wooler via Tosson, Thropton, Alnham, Glanton and Wooperton. Alnwick businessmen thought little of this idea to divert business from Alnwick and persuaded the North Eastern Railway to build a line from Alnwick to Edlingham, Whittingham, Wooler and on to Cornhill on the south side of the Tweed, opposite Coldstream. This latter proposal won the day, an Act of Parliament was passed in 1882 and the line was opened in 1887 to serve that part of mid-Northumberland that lies between the Cheviot Hills and the main east coast line from Newcastle and Edinburgh. No doubt the NER were influenced in constructing this line by the probability that if they did not do it, the North British would.

The Wooler line was 36 miles long and ran from a terminal station at Alnwick, itself on a branch line from Alnmouth, to Wooler and beyond to join the North Eastern Railway's branch line from Tweedmouth to Kelso at Cornhill, or Coldstream. The same station was given both names during its lifetime.

Leaving Alnwick in a south-westerley direction, the line turned north-west to climb over Alnwick Moor at a height of 620 ft with gradients as steep as 1 in 50. Passing under the Alnwick to Rothbury road (B6341), the railway then descended to 250 ft to cross the Edlingham Burn on an arched stone viaduct. The only tunnel on the line was at Hillhead, between Edlingham and Whittingham, with a length of 350 yards. From Whittingham station to Wooler the railway ran through a broad glacial valley that improved in fertility as it went north. The alignment was roughly parallel to the A697 road and, like the road, it crossed numerous small burns and rivers that ultimately form the River Till which in turn flows into the River Tweed at Twizell. At Wooler the line turned west up the valley of the River Glen and then through a gap cut by the Bowmont Water through a northern spur of the Cheviots. Bowmont Water was crossed at Mindrum and the railway continued north to join the Tweedmouth-Kelso line south of Coldstream station.

The entire route was through agricultural country devoid of any mineral or industrial activity. Farming varied from livestock rearing with suckler cows and upland sheep at the southern end of the line to progressively improving agriculture in the valley of the River Till and some of the best farming land in the country on Tweedside. Here, grain (especially malting barley) and potatoes competed with the production of prime beef and fat lambs. Goods traffic, therefore, consisted of livestock, including hunters and bloodstock horses, and grain being taken out of the area and agricultural requisites such as fertiliser, feeding stuffs and machinery being brought in. If goods traffic was light, passenger traffic was even lighter.

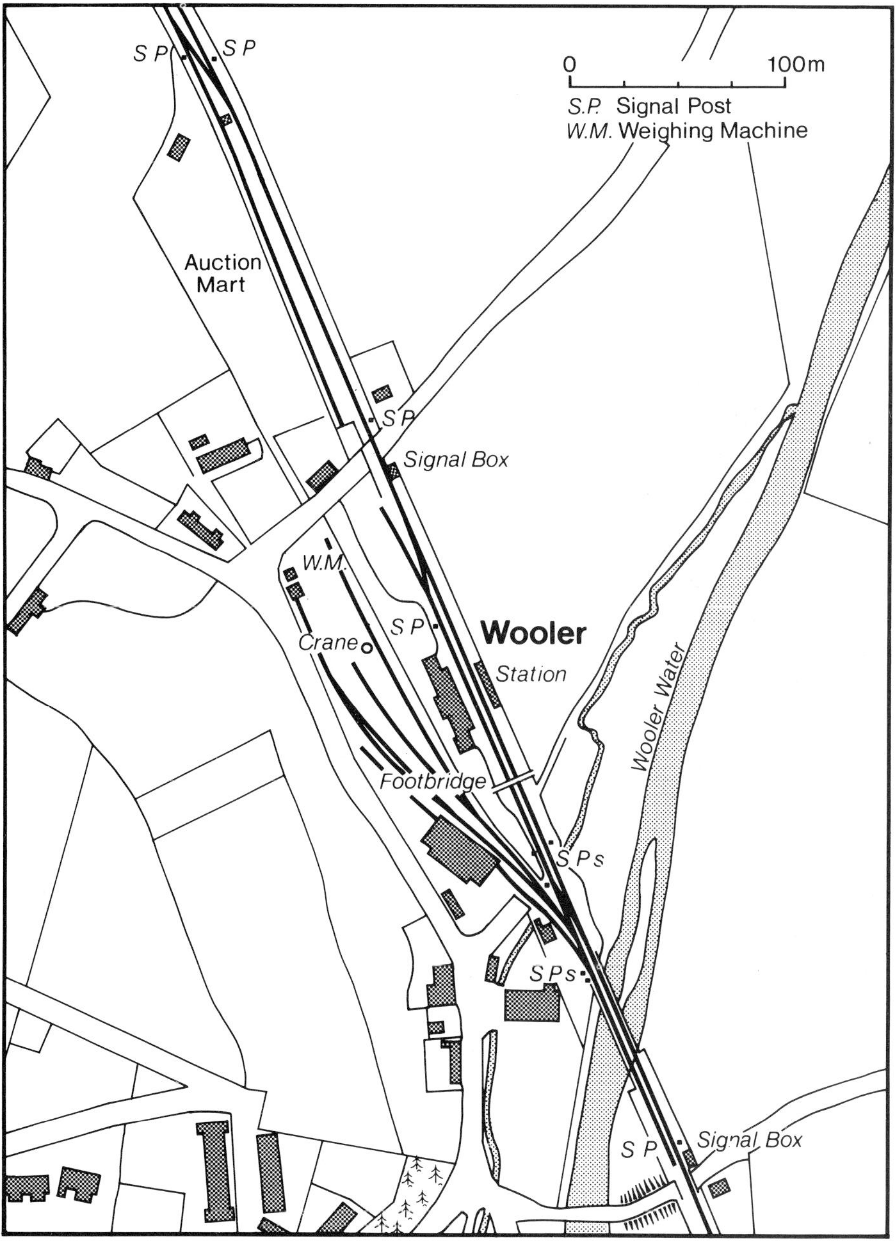

Wooler station

At the peak of the line's activity, passenger trains never exceeded three a day in each direction and there were none on Sundays. Surprisingly, the service was run from the Coldstream end of the line with the first train of the day leaving there at 6-38 a.m. This, in fact, was a through train which left Berwick at 5-55 a.m., though it wasn't advertised as such. It arrived at Alnwick at 8-10 a.m. and departed again for Wooler and Coldstream at 8.27 a.m. It then had 40 minutes at Coldstream before leaving at 10-40 a.m. to arrive at Alnwick at 12-16 p.m. The afternoon train left Alnwick at 1-50 p.m., called at Wooler at 2-51 p.m. (an hour for 22½ miles) and arrived at Coldstream at 3-23 p.m. (another 32 minutes for 13½ miles!). The final train of the day departed from Coldstream at 4-29 p.m., arrived at Alnwick at 6-03 p.m., left again at 6-42 p.m., got to Coldstream at 8-15 p.m. and terminated at Tweedmouth at 8-42 p.m., though again not advertised as a through train.

The advent of the motor-bus in the 1920s, with its ability to go into villages by-passed by the railway, sealed the doom of passenger services and they were withdrawn in 1930. This left a series of very attractive stone-built stations at Edlingham, Glanton, Hedgeley, Wooperton, Ilderton, Wooler, Akeld, Kirknewton and Mindrum and many of them have been converted into desirable homes. The exception is Whittingham which was built as an island platform, the only one on the line, and has remained derelict ever since. The glass in the canopy over the platforms remained in situ long after the track was lifted but on my last visit, some years ago, it had all been broken and only the gaunt ironwork remained.

The station at Wooler, like all the others on the line, was designed by William Bell and featured the half-hipped roofs for which he was noted. It is a substantial two storey stone building on the west, or down, platform. It included two waiting rooms, booking office, luggage office and station master's house. A glass canopy covered about half the width of the platform for the full length of the building. There was also a small glazed timber waiting room on the east, or up, platform.

The goods yard contained four sidings one of which ran through a large stone-built goods shed with covered loading doors for road vehicles. Another siding ran to a batch of typical North Eastern Railway coal cells. The goods yard was worked from a long headshunt, at the south end of the station, which ran right up to but did not cross Chatton Road.

In 1936 the main station building was taken over by the Youth Hostels Association and their volunteers prepared it for its new role. One waiting room became a female dormitory and the other a male one. Sexes were positively separated in those days. The ticket office was transformed into the kitchen. The Y.H.A. ceased to operate during the war and the premises became a hostel for the Women's Land Army only to return back to Y.H.A. use once the war was over.

In 1935 the regular service consisted of a daily goods train from Alnwick to Wooler and return, a daily goods train from Coldstream to Wooler and return, and a daily No. 1 Express Parcels train (to quote the working time-table) which

ALNWICK, KELSO and COLDSTREAM to BERWICK—Weekdays

Distance from Alnwick	Distance from Kelso		1	4	6	10		12	14	16	18	22		24		26	27	28	30	31		32	34		36	37	38	42	SUNDAYS 1
				A	D	B				B	D	B					D		D	D		A			B		B		
			RAIL MOTOR	Goods	Goods	Cattle		PASSENGER	RAIL MOTOR	Goods	Goods	Cattle		No. 1 EXPRESS PARCELS		PASSENGER	Goods	Light Engine	Cattle	Goods		Cattle	PASSENGER		Cattle		Cattle	RAIL MOTOR	EXPRESS PASSENGER
						FOQ				SX	SO	MO					SX	SX	Q	SO		Q			Q		Q	SO	
						arr.	dep.					arr.	dep.	arr.	dep.					arr.	dep.		arr.	dep.					
			HC	K	P			F	HC	PJ	P							V										HC	S
M.C.	M.C.		a.m.	a.m.	a.m.	a.m.	a.m.	a.m.	p.m.	p.m.	p.m.	p.m.	p.m.	p.m.	p.m.	p.m.	p.m.	p.m.	p.m.	p.m.	p.m.	p.m.	p.m.	p.m.	p.m.		p.m.	p.m.	p.m.
DOWN	—	**ALNWICK** .... ⊕	—	—	*	—	—	—	—	—	—	—	—	—	1 48	—	—	—	—	—	—	—			—		—	—	—
4 26	—	*Summit*	—	—	b	—	—	—	—	—	—	—	—	..	..	—	—	—	—	—	—	—	—	—	—		—	—	—
7 0	—	Edlingham	—	—	*	—	—	—	—			—	—	....	..	—		—	—	—	—	—			—		—	—	—
9 70	—	Whittingham .. ⊕	—	—	*	—	—	—	—	Return of No. 11 Up	Return of No. 11 Up	—	—	2 2	2 5	—	Return of No. 19 Down	—	—	—	—	—			—		—	—	—
11 47	—	Glanton	—	—	*	—	—	—	—			—	—	.....	2 10	—		—	—	—	—	—			—		—	—	—
13 44	—	Hedgeley .. .. ⊕	—	—	*	—	—	—	—			—	—	..	2 15	—		—	—	—	—	—	St. Boswells	dep. 5.57 p.m.	—		—	—	
15 57	—	Wooperton	—	—	*	—	—	—	—			—	—	.....	2 21	—		—	—	—	—	—			—		—	—	
18 49	—	Ilderton	—	—	*	—	—	—	—	—	—	—	—	..	2 28	—		—	—	—	—	—			—		—	—	
22 13	—	Wooler .... ⊕	—	—	DD	—	—	—	—	U	U	—	1 45	2 35	2 38	—		—	6 0	—	—	—			7 0		9 0	—	
24 63	—	Akeld .. .. .. ⊕	—	—	—	—	—	—	—	*	*	1 55	2 5	..	2 43	—		—	..	—	—	—			..		..	—	
27 59	—	Kirknewton.........	—	—		—	—	—	—	*	*	....	*	.....	2 50	—		—	.....	—	—	—			.....		.....	—	
30 8	—	*Kilham Siding* ..	—	—		—	—	—	—	*	*	..	..	..	..	—		—	..	—	—	—			..		..	—	
32 8	—	Mindrum	—	—		—	—	—	—	*	*	2 25	2 35	.....	2 58	—	—	—	.....	—	—	—	—	—	.....		.....	—	—
—	—	**KELSO**	—	U	Alnmouth dep. 8.0 a.m., p. 8	—	—	10 0	12 32	—	—	—	—	—	—	3 5	U	—	—	—	5 55	6 30	6 18	6 38	—		—	—	7 20
UP —	2 22	Sprouston	—	**		—	—	10 5	12 37	—	—	—	—	—	—	3 10	*	—	—	6 2	6 13	....	.....	6 43	—		—	—	.....
—	4 46	Carham	—	**		—	—	10 10	12 42	—	—	—	—	—	—	3 15	*	—	—	..	*	..	..	6 48	—		—	—	..
—	6 67	Sunilaws	—	**		—	9 50	10 15	12 47	—	—	—	—	—	—	3 20	*	—	—	.....	*	*	.....	6 53	—		—	10 23	.....
35 55	9 72	**COLDSTREAM** . arr.	—	*		10 0	..	10 20	..	—	—	2 45	..	3 5	—	3 25	..	—	6 40	6 30		6 50	6 58	..	7 40		9 40	..	..
		dep.	8 13			.....	10 30	10 23	12 53	*	*	.....	3 5	—	—	3 27	*	4 41	7 5	.....	7 10	7 16	....	6 59	7 49		9 49	10 29	7 34
38 48	12 65	Twizell	8 19	To Wooler		10 40	10 48	10 29	12 59	..	*	..	*	—	—	3 33	*	..	..	..	*	..	..	7 5	..		..	..	..
41 38	15 55	Norham	8 25			10 58	11 5	10 35	1 5	.....	*	3 25	3 50	—	—	3 39	*	....	.....	.....	*	.....	...	7 11	...		..	.....	7 42
44 1	18 18	Velvet Hall	8 30			11 15	11 25	10 40	1 10	..	*	4 0	4 10	—	—	3 44	*	..	..	7 40	7 55	..	..	7 16	..		..	..	..
48 11	22 28	**Tweedmouth** arr.	8 37	p. 3		11 40	—	10 47	1 17	DD	D	4 25	—	—	—	3 51	DD	4 59	7 38	8 10	—	7 41	7 23	.....	8 20		10 20	10 48	8 0
		dep.	8 40	—	—	—	—	10 52	1 19	—	—	—	—	—	—	3x59	—	—	—	—	—	—	..	7y33	—		—	10 53	8 7
49 23	23 40	**BERWICK**	8 43	—	—	—	—	10 55	1 22	—	—	—	—	—	—	4x2	—	—	—	—	—	—	7y36	—	—		—	10 56	—

Sunday column 1: SUSPENDED.

F—"HC" except from or to one intermediate station by pre-arrangement with D.P.M. J—Attaches live stock at Twizell, Norham and Velvet Hall when required.
K—Shunts Sprouston Siding. Worked by trainmen of No. 1 Down. p. 3. P—Conveys parcels, etc. traffic, also horse boxes and carriage trucks. S—Newcastle **arr.** 9.38 p.m., p. 27
V—**Runs to 16th December and from 9th January to 31st March.** Z—Steam train SO. b—Brakes. x—When No. 220 p. 11 runs, departs Tweedmouth 3.56 p.m. and arrives Berwick 3.59 p.m.
y—Unless train can be despatched at booked time must depart Tweedmouth 7.45 p.m., Berwick arrive 7.48 p.m. **—No. 4 Stops when required for live stock only.

A 30 m.p.h. B 25 m.p.h. D 20 m.p.h. O Run when required U Stops only to take up or attach

From the L.N.E.R. (North Eastern Area) "Working Timetable", September 1938

Photo: Stuart Sellar

Ivatt Class 2 2-6-0 No. 46482 standing at Akeld station 29th March 1965. This was the railhead for R.A.F. Millfield during the war.

Photo: Stuart Sellar

Ivatt Class 2 2-6-0 No. 46482 at Yeavering Crossing 29th March 1965. Even gate boxes were of substantial construction!

left Alnwick at 1-40 p.m., stopped at all stations except Edlingham, arived at Wooler at 2-35 p.m. and Coldstream at 3-05 p.m. The return working departed from Coldstream at 4-11 p.m., called at Wooler at 4-43 p.m. and arrived at Alnwick at 5.42 p.m. This train called at all stations, including Edlingham, which is no doubt why it was given the title of No. 1 Express Parcels! Consignments for delivery in the Wooler area were distributed by green painted Country Service delivery vans.

One of the features of our family holidays at Wooler was that they always included my grandmother. She was the widow of an ex-LNER railwayman and, as such, she was entitled to privilege tickets. She made sure that she used them for her method of travel to Wooler was to take the train from Tynemouth to Alnwick and then the United bus from Alnwick to Wooler. She could just as easily have gone from Newcastle to Wooler direct by bus, probably more quickly, but that would have prevented her from using her P.T. On arrival at Alnwick station, grandmother transferred to the bus which called at the station, but her luggage was put on to the No. 1 Express Parcels for Wooler. We collected it from Wooler station any time after the train's arrival at 2.35 p.m.

On these occasions we arranged for a lunch-time picnic at a point on the B6341 Alnwick to Rothbury road where it ran close to the railway. The importance of the picnic was not the food but the almost silent appearance of the No. 1 Express Parcels as it drifted out of the cutting from Alnwick Moor summit down the gradient to Edlingham, cross the viaduct over Edlingham Burn and disappear into Hillhead tunnel. We had a personal interest in the safe progress of this train because it was carrying grandmother's luggage.

In addition to the daily goods and parcel trains, there was a regular weekly cattle train on Mondays to take fat cattle to Tweedmouth for onward journeys to Edinburgh and Tyneside. This train was booked to leave Wooler at 1-45 pm. and run to Coldstream ahead of the parcel train. The busy time of the year for livestock was from August to October. Lamb sales commenced at Wooler auction mart in August. These were followed by draft ewe sales and then by suckled calf sales which ran into October. Extra trains were run as required to deal with this special seasonal traffic and provision was made in the time-table for depatures from Wooler at 6-00 p.m., 7-00 p.m. and 9-00 p.m. – all to Tweedmouth for forwarding. Wooler mart also sold considerable numbers of imported Irish store cattle in the spring of the year. These were brought into Wooler by train from the port of arrival via Coldstream as and when required.

Akeld station, three miles north of Wooler, became very important in the 1940s when it was the railhead for a new R.A.F. fighter aerodrome at Milfield (four miles to the north). Subsequently, after the end of hostilities, some of the hanger space was used for food storage and, again, Akeld station was the rail-head for this traffic.

Back in 1935, Wooler station reached its daily peak of railway activity between 10-30 a.m. and noon. First to arrive was the goods train from Alnwick heralded by the opening (to the train) of the level-crossing gates on the Wooler to Chatton road at the south end of the station. On arrival at the west, or down, platform the engine would be detached, run forward, then back through the up platform to work the train from the rear. The goods shed and coal staithes were behind the down platform with access from the Alnwick end of the station (see station diagram). Shunting operations inevitably fouled the level crossing so that the gates frequently had to be opened and closed, much to the annoyance of traffic on Chatton road, even in those days. There was also an oil/petrol depot belonging to Shell and B.P. on the Alnwick side of Chatton road with access by facing points from the station.

While shunting was proceeding at a fairly leisurely pace, the goods train from Coldstream would arrive, having first been from Coldstream to Kelso and back. This arrival was similarly heralded by the opening of the crossing gates on the Wooler to Berwick road at the north end of the station. The next hour or so was pure ecstasy for a small boy. Two engines shunted and took it in turns to detach, attach, and fly-shunt wagons and vans from both trains into and out of the goods yard. If there were any empty cattle wagons, these had to be propelled into the auction mart siding on the north side of the Berwick road.

Eventually, sometime between noon and 12-30 p.m. (I was always being scolded for being late for our twelve o'clock dinner!) all the sorting would have been completed, two new trains made up, and each would depart from whence it had come, one to Alnwick and the other to Coldstream. The hectic activity was over for another day and Wooler station resumed its tranquil existence until the arrival of the down No. 1 Express Parcels at 2-35 p.m.

In the 1930s the Coldstream branch was supplied with engines from both Alnmouth and Tweedmouth sheds. Alnmouth provided the engine for the morning goods from Alnwick and for the afternoon parcel train. Tweedmouth supplied the engine for the goods train from Coldstream and for special livestock trains. The morning goods began at Tweedmouth at 6-40 a.m. and worked all stations to Kelso. It then returned to Coldstream, shunting Sprouston siding "en route", and worked to Wooler calling at Mindrum, Kilham sidings, Kirknewton and Akeld as required. The return trip was to Coldstream and direct to Tweedmouth on Mondays to Fridays, but on Saturdays it called at Twizell, Norham and Velvet Hall as required.

There was no turn-table at Wooler so tank engines were usually used for the goods trains. Time has erased from memory the classes of locomotive involved but a study of those stabled at Alnmouth in those days suggests that either G5 0-4-4's or N8 or N10 0-6-2's were the most likely. Gradients between Alnwick and Whittingham were such that G5's were limited to loads not exceeding 145 tons so it was probably an N8 or an N10 that handled the average goods train of 12 to 14 wagons. A correspondent has told me that in 1940 a J21 0-6-0 tender engine was replaced by an N10 tank.

Alnmouth, at that time, was also a sort of retirement home for elderly main-line passenger locomotives. One of these usually had the gentle task of hauling the solitary bogie parcels van from Alnwick to Coldstream and back. An ex-North Eastern "R" class (LNER class D20) 4-4-0 was a regular choice for this turn. Tweedmouth shed also had an allocation of G5's, N8's and N10's and, since

I have no recollection of seeing tender engines running tender first, it is likely that goods trains from Coldstream were hauled by tank engines of these classes.

Even though traffic was sparse and the number of trains was limited to those described, full block signalling was maintained after the withdrawal of passenger services and right up to the end of LNER control in December 1947. Whittingham, Hedgeley, Wooler and Akeld were block posts with passing loops. Wooler had two signal boxes but the one at the south end of the station was only a gate box for the level crossing on Chatton Road. At all the other stations access to sidings was by ground frame controlled by either electric tablet or staff or by a key attached to it. Level crossings, and there were eight of them, were operated either from adjacent signal boxes or by crossing keepers who lived in line-side cottages provided with a frame to operate protecting signals.

Signalling arrangements throughout the line are summarised as follows:-

Miles from Alnwick	*Location*	*Apparatus*
6.8	Edlingham station	Ground frame
9.7	Whittingham station	Signal box, block post
11.4	Glanton station	Ground frame
13.4	Hedgeley station	Signal box, block post
15.6	Wooperton station	Ground frame
18.5	Ilderton station	Ground frame
22.1	Wooler station	Two signal boxes, block post
24.7	Akeld station	Signal box, block post
27.6	Kirknewton station	Ground frame
29.3	Kilham siding	Ground frame
31.4	Mindrum station	Ground frame
35.7	Coldstream station	Signal box, block post

Alas, full running of the line ceased shortly after nationalisation because the line was breached when the bridge across the Lil Burn, immediately to the north of Ilderton station, was washed away in a torrential thunderstorm in August 1948. The cost of repair was out of all proportion to the likely forthcoming revenue so, from then on, the southern section of the line was worked from Alnwick to Ilderton. The absence of a passing loop at Ilderton meant that all trains had to be propelled back to Hedgeley before the engine could run round the train. Not the best of ways in which to run a railway. This section of the line was closed in March 1953.

The northern section from Coldstream to Wooler continued to operate as before the wash-out until November 1957 when Wooler North signal box was closed and the section from Akeld to Wooler was operated as "one engine in steam". The Akeld to Coldstream section continued as a token block until the line closed. Meanwhile, Wooler continued to accept agricultural requisites of feed, machinery, fertilisers, oil and coal until the end and dispatched considerable numbers of sheep and cattle. A useful addition to traffic during the last few years was the seasonal sugar-beet crop. This was taken by train from stations north of Wooler to the British Sugar Corporation factory at Cupar in Fife. This factory was so short of tonnage from Scottish sugar-beet growers that not only did it accept crops from Northumberland and Durham but it also paid for the carriage of the crop to the factory by train.

The end came with the Beeching axe of 1965 when the entire Tweed valley was deprived of railway services. Perhaps it is significant that, as sugar-beet is no longer grown in the North of England, both sugar factory and railway are now extinct.

The Wooler line was a delightful rural railway that had many scenic attractions but the LNER was reluctant to provide opportunities for the public to enjoy them. I have been informed that there were a few scenic excursions in the late 1930s of which I was hitherto unaware. These were run from Newcastle to Tweedmouth, then to Coldstream and Wooler where there was a break for tea. Trains were made up of teak tourist coaches and included a Buffet car. Motive power was a J39 0-6-0. A D20 4-4-0 was sent out light engine from Alnmouth to Wooler to assist the returning train on the climb from Whittingham to the 620 ft summit on Alnwick Moor. Had I known of these excursions my ambition could have been fulfilled. Sadly, they were not repeated after the war and now, of course, the line has gone.

In Retrospect

In collecting and assembling material for this book, many long-forgotten memories have been recalled. The verification of these has brought about the renewal of old friendships and the creation of many new ones. I am greatly indebted to all who have helped, not only by confirming my own memories, but by contributing information concerning their own recollections and experiences of Northumberland's railways.

My great regret is that I did not take any photographs of these railways when they were alive and active. Although a camera was available, it was not the custom or practice to take pictures of trains. Why should anyone want to do such a silly thing with expensive film? Railways were a permanent feature of the landscape, they would be there for ever, they could be seen in action at any time. Why take photographs? Railways were part of life and their operations were expected, accepted and taken for granted as an essential ingredient of life in rural Northumberland. Then, suddenly, they had gone.

Fortunately, there were a few dedicated people who had the foresight to do silly things like taking photographs of railways and trains. Without them this book would have been devoid of pictures. Many of these photographers have now passed on but their works remain as a permanent record of what rural Northumberland used to be like. To them, and to those who are still with us, I say "Thank you" for your foresight and tenacity in providing the rest of us with such a wonderful visual record of railways that are now history.

It is also a matter of regret that almost all of the railways recalled in this book were closed before any organised groups of people challenged such closures and sought to have them developed as tourist attractions. It all happened at a time when the railway preservation movement was in its infancy for, although the Talyllyn Railway Preservation Society succeeded in keeping that line open after the death of its owner, Sir Henry Haydn Jones, in 1950, the Bluebell Railway was the first standard gauge line to be preserved and it re-opened in 1960. By that time the Allendale, Border Counties and Wooler lines had gone.

The one exception is the Alston line where the sounds of steam can still be heard coming from 1½ miles of track, albeit 2 ft gauge, running north from

Photo Richard Graham

Above: As abandoned by British Rail
Below: As revived by the South Tynedale Railway Preservation Society

Photo: Allan W. Stobbs

Photo: Allan W. Stobbs

Above: No. 10 "Naklo" Polish-built 0-6-0 near mile post 1 north of Alston
Below: No. 6 "Thomas Edmondson" German-built 0-4-0 heads south over the Tyne Bridge

Photo: Allan W. Stobbs

Alston station. When British Railways (as it then was) finally closed the line in May 1976 the South Tynedale Railway Preservation Society was formed to take over the abandoned track and run trains on at least part of it. However, the asking price for scrap rail and sleepers was so absurdly high that this proposal came to nought. Instead, and with help from Cumbria and Northumberland County Councils who bought the track-bed and stations, the Society forged ahead with a 2 ft gauge line which opened to traffic in July 1983. This initial mile of track was extended another half mile to the county boundary at Gilderdale in December 1986. Since then, trains have been running regularly at weekends and daily during July and August. Generally, steam locomotives are in action on Saturdays, Sundays and Bank Holidays. All other services are diesel hauled. Passenger figures have gradually built up to 25,000 a year. The railway owns 3 steam engines, two German and one Polish, and has the use of two others. There are also a number of diesel locomotives. Pride of place on the railway is the ex-North Eastern Railway signal box from Ainderby on the Wensleydale line which has been painstakingly re-erected at Alston station. It contains a 21-lever McKenzie and Holland frame from Leaming and the gate-wheel from How Mill, which now operates lifting barriers. The line is gradually being extended and is now (1992) across Lort Burn Viaduct and heading for the hamlet of Kirkhaugh. Hopefully, Slaggyford will be reached in the not too distant future.

Whether or not the growth of tourism in Northumberland has been helped or hindered by the demise of the railways I do not know. The success of the Dales Rail system on the Settle-Carlisle railway over a period of 10 years and the essential role that it played in the revitalisation of the line indicates that there are large numbers of people in urban areas who are only too happy to leave their cars at home, or their local station, and use trains to explore and enjoy the countryside. The Border Counties line could have played a significant role in the development of Kielder, both Forest and Water, as a tourist centre – and brought out the timber as well.

Undoubtedly, the growth of road transport, both private and commercial, killed-off these rural railways. It is ironic to learn that, in the 12 year period since Dr. Beeching wielded his axe, the total savings from all the line closures amounted to no more than what it cost to build three miles of urban motorway. What, then, has been achieved?

We live in a world of change and what was acceptable in, say, 1943, is not acceptable today. But "change" and "progress" are not synonymous. Changes today take place all too often because of vested interests and minority pressure groups and not for the benefit of the community as a whole. The ordinary person is never even asked. It is questionable whether the abandonment of rural railways in Northumberland has increased the quality of life in the county. It may have saved some money but where has it gone and who has benefitted from the savings?

The owner of a multi-thousand pound K-registered car pays at least 30p a mile to keep it on the road. Is he happier or more content than the traveller waiting on Bellingham station platform for the 5-36 p.m. to Kielder at 1d a mile Third Class? Alas, he no longer has the choice.

Photo: Allan W. Stobbs

"The descent into oblivion"
The 4-48 p.m. from Riccarton pulling away from Saughtree behind an ex-North British 4-4-0, LNER Class D30.
A case of "Down and out".

Further Reading

Rural Branch Lines of Northumberland — C. R. Warn
published by Frank Graham 1975

Forgotten Railways of North-east England — K. Hoole
published by David & Charles 1973

North Eastern Locomotive Sheds — K. Hoole
published by David & Charles

Forgotten Railways of Scotland — John Thomas
published by David & Charles 1976

The Rothbury Branch — Stanley C. Jenkins
published by The Oakwood Press 1991

The Newcastle & Carlisle Railway — G. Whittle
published by David & Charles 1979

Branch Line Trains in the North-east — K. Hoole
published by Dalesman 1979

North Eastern Stations — K. Hoole
published by Dalesman 1978

The Alston Branch — Stanley C. Jenkins
published by The Oakwood Press 1991

Industrial Archaeology of North-East England — Frank Atkinson
published by David & Charles 1974

Two Centuries of Industrial Welfare — Arthur Raistrick
published by Moorland Publishing Company 1977

Life & Work of the Northern Lead Miner — Arthur Raistrick & Arthur Roberts
published by Alan Sutton Publishing 1990

Acknowledgements

The revision and enlargement of this book has given me a great deal of pleasure, not least because I have come to know so many kind-hearted enthusiasts who have been so willing to provide photographs, time-tables and accounts of their own experiences.

I have already made reference to the contributions made by Ken Appleby, J. B. Dawson, Bill Lynn, and Dr. J. L. Middlemas and to these I would like to add Mr. P. McKenzie of Vickers Defence Systems for information on "The Ridsdale Ironstone Workings".

The provision of suitable illustrations is always a problem and I am greatly indebted to Ian S. Carr and Bill Lynn for making available such an excellent supply of prints. Other contributions have come from Alex Appleton, John Daysh, Richard Graham, Jim Foster-Smith, Lens of Sutton, Stuart Sellar, Ken Taylor, and Robin Walton to all of whom I express my thanks.

My special thanks go to Peter Thorn, G.R.A., for his pencil drawing of G5 1918 at Rothbury, which graces the front cover.

Mrs. Susannah Haley has added three new diagrams to those she did for the original publication and the standard of detail and clarity is as good as ever.

Ian Edgar has devoted a lot of time and skill in improving the presentation of old negatives and prints.

Malcolm Warwick and Derick Woods of Reed's Printers have shown more than the usual interest in the printing and presentation of this book and their advice and help in formulating the lay-out has been invaluable.

To all of these people I say "Thank you".

Finally, a special word of thanks to my wife who suffered my long absences in the study and then so painstakingly checked the final typescript.